# Marangoni Convection in Microgravity: Unveiling the Secrets of Non-Periodic Flow

Chopra

## TABLE OF CONTENTS

# 1  INTRODUCTION

Temperature gradients in liquid-liquid and liquid-gas systems evolve both free and forced convection influencing heat and mass transfer processes within the system. On earth, the convective flow due to buoyancy is the dominant flow, which is a result of density variations due to temperature gradients in the flow domain. Moreover, a non-neglected secondary convective flow is observed on the phase boundary between two liquids or liquid and gas. This additional convective flow is a result of temperature gradients at the phase boundary. The interfacial tension variation with temperature ($\partial\sigma/\partial T$) caused by interfacial temperature gradients ($|\partial T/\partial z|$) leads to a fluid flow towards regions of high surface tensions and low temperatures, well-known as thermocapillary or Marangoni convection. The physical problem is mainly characterized by a dimensionless number that represents the ratio of convective heat flow induced by capillary convection to the heat transfer due to conduction named Marangoni number (Mg).

Due to its influence in different heat and mass transfer processes as well as material manufacturing in space, Marangoni convection is widely considered to be more critical and important under microgravity conditions. It can dominate heat and mass transfer processes. Therefore, manufacturers should consider thermocapillary convection in material production processes in space [1]. Moreover, one can purposefully use the temperature gradient to eliminate or move bubbles or drops suspended in liquid metals [2], [3], as well as in other applications like manufacturing of single-walled carbon nanotubes [4], mono crystal production, and investigating flame spread above liquid fuels [5], to mention only a few examples.

Researchers have always seen Marangoni convection as an interesting topic for both numerical and experimental studies. The past decade has seen different approaches to describe the flow behaviour under higher Marangoni numbers, where the flow turns from its stable laminar behaviour into a non-laminar one characterized by periodic or non-periodic oscillations. Previous studies have focused on critical values of Marangoni numbers for different fluids where periodic or non-periodic oscillations start to appear, termed the critical Marangoni number ($Mg_c$). However, they have not yet presented a detailed image of the non-periodic oscillatory flow at higher temperature gradients. Therefore, the major aim of the current study is to identify different factors affecting the non-periodicity of Marangoni flow and to explain how this flow behaves under conditions beyond the critical Marangoni number.

In this study, thermocapillary-driven convection around a bubble under a heated wall is experimentally investigated under gravitational conditions as shown in figure 1. The induced convective Marangoni flow is categorized into laminar, periodic oscillatory, and nonperiodic oscillatory flow through the patterns of the tracer particle motion around the bubble using particle image velocimetry (PIV), through surface temperature variation measurements using resistance temperature detectors (RTDs), and through the transient behaviour of the thermocapillary vortex boundary contour using shadowgraphy. Moreover, further experiments were pursued under gauge pressure $\Delta p$ using an appropriate pressure chamber. The implementation of the mentioned experiments under gauge pressure conditions formed a novel methodology permitting to do such experiments under higher temperature gradients. The overall structure of the presented book includes four chapters. Chapter two provides a brief overview of the recent history of Marangoni convection research and its characteristic dimensionless numbers and explains how past authors contributed with both experimental and numerical studies. The third chapter is concerned with the methodology used for this study and the experimental setup. The fourth section presents the results of experimental work, focusing on different experiments under normal or gauge pressure conditions. In the last chapter the results are analyzed and discussed in comparison with previous studies. Moreover, conclusions and suggestions for future possible experimental and numerical work are given.

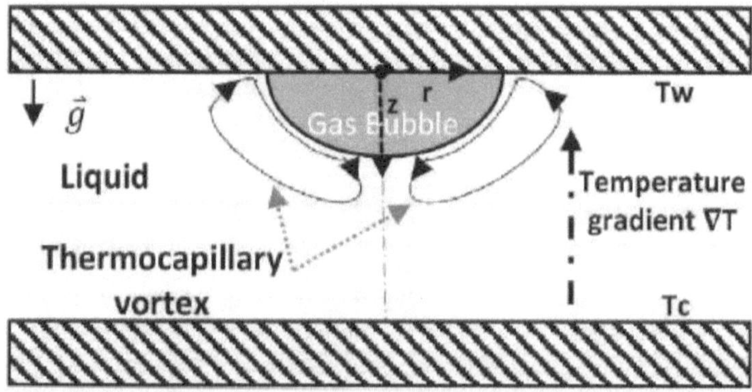


*Figure 1.1: Schematic of thermocapillary bubble convection under a heated wall*

# 2 LITERATURE REVIEW

## 2.1 Preface

The interfacial flow caused by surface tension gradients, later termed Marangoni convection, was formerly identified and studied in 1855 by James Thomson [6]. He pointed out that varying the concentrations of different liquids and thus their different surface tensions can cause an internal liquid flow. He explained correctly the known "tears of wine" phenomenon. These observable motions of alcohol droplets on a wine glass were attributed to the concentration gradients of the wine on the glass. Later, in 1877, the Italian scientist Carlo Giuseppe Matteo Marangoni [7], whom the mentioned convective flow was named after, studied the effect of scattering one liquid over a second immiscible liquid. He indicated that the first liquid can spread over the second one only if the sum of interfacial tension and the surface tension of the first liquid is lower than the surface tension of the second one. In 1900, the convective thermocapillary flow in thin liquid layers above heated surfaces was observed by Bénard [8]. Later thereafter, it was concluded that those convective cells, which were discussed by Rayleigh [9], were also affected by Marangoni convection [10]. Since that time, Marangoni convection evolved into an interesting research topic, and the contributions of early scientists were later reviewed by others including Nempomnyashchy et al. [11].

Research study on Marangoni convection has extended to include many different sub-topics like: hydrodynamic instabilities due to thermo-solutal Marangoni effect, Bénard-Rayleigh convection and its evolution, the ability of thermal gradients to eliminate and move bubbles, the effect of induced thermocapillary convection on the evaporation process of different volatile liquids, the condensation and evaporation processes on thin liquid layers subjected to free surfaces, Marangoni convective flow effects on liquid fuel atomization and combustion, and many other chemical processes. The mentioned examples include complex flows, where thermocapillary convection plays an important role. For example, many authors conducted a couple of experimental and numerical studies to understand the influence of thermocapillary convection on boiling heat transfer [12]–[16].

To mention only few examples, G. Wozniak et al. [17] presented a comprehensive review regarding the thermocapillary convection driving bubbles and drops including many experimental and theoretical studies. The aim of that study was to summarize the previous effort and to give information about the arising research questions regarding that topic like: theoretical modelling and experimental verification of thermocapillary fluid particle motion at higher Reynolds and Marangoni numbers, and theoretical and experimental studies on the problem of fluid particle interaction with other fluid particles and solid walls.

Moreover, Li [18] presented a comprehensive review regarding thermocapillary instabilities in both simple and binary fluids. In addition to that he conducted experiments on Marangoni convection in a thin liquid layer driven by a horizontal temperature gradient in a sealed rectangular cavity for a simple fluid. In his analysis, most of the evaporation and condensation was observed to occur at the liquid-vapour interface and near the heated and cooled ends, so that the temperature gradient was parallel to the liquid-vapour interface. He showed how the adverse effects of thermocapillarity in both condensers and evaporators in simple coolants can be reduced using binary fluid coolants like water-alcohol mixtures. For example, in a methanol-water mixture, the resulting solutocapillary stresses can effectively oppose the thermocapillary stresses. The effect of the presence of any noncondensables (i.e., air) in this sealed system, which often occurs in practice, was also presented. It formed a ternary mixture of air and the vapours of the two components in the vapour space above the binary-liquid layer. The air concentration in the vapour space above the liquid layer was found to have a significant effect on the complicity of the flow behaviour.

In general, much effort was done to review thermocapillary instabilities, which were due to external imposed temperature gradients normal or parallel to the interface or the free surface [19].

It is nearly impossible to give a complete review on Marangoni convection literature especially for the work which included different flow complexes at the same time as: buoyant convective flow, evaporation, condensation, soluto-capillary flow, and thermocapillary convection. The aim of this work, as mentioned before, is to investigate to a large extend undisturbed thermocapillary convective flow and to find the factors affecting the instabilities accompanied by that flow. The investigation of pure thermocapillary flow can only be achieved under microgravity conditions, where the thermocapillary flow is the dominant one. Therefore, we will focus only on studies related to thermocapillary convection induced around a stationary bubble under normal and microgravity conditions as configuration to demonstrate, discuss and analyze thermo-convective liquid flow. Other typical experimental configurations as rectangular or cylindrical test cells with flat or curved liquid surfaces, including rotation or liquid bridges, were investigated by Szymczyk et al. [20] and Velten et al. [21].

## 2.2 Definition of the Marangoni Number

According to the mentioned experimental setup of a bubble injected under a heated wall, with the presence of temperature and surface tension gradients along the air-liquid interface, the buoyancy force near the interface is counter-acting the Marangoni force. Compared to "pure" thermocapillary flow under zero gravity conditions, a slow-down of the coupled convection near the bubble interface is observed due to this coupling effect of the two driving forces. To define the conditions of this flow configuration, two dimensionless groups are defined: a Marangoni number and a bubble aspect ratio ($\frac{r_B}{z_B}$), which is defined as the ratio between the injected bubble maximum horizontal radius ($r_B$) and its vertical expansion ($z_B$). The term Mg-number is generally understood as the ratio of convective heat transfer due to a gradient of surface tension to the heat transfer due to thermal diffusion. The Marangoni-number can be expressed as the product of the Reynolds and the Prandtl-number,

$$Mg = Re\ Pr \qquad (2.1)$$

or

$$Mg = \frac{l\ U}{\nu}\frac{\nu}{\alpha} \qquad (2.2)$$

where $Re = \frac{l\ U}{\nu}$ and $Pr = \frac{\nu}{\alpha}$.

Here, $l$ denotes a characteristic length in the Reynolds number ($Re$), $\nu$ denotes kinematic viscosity, and $\alpha$ the thermal diffusivity of the test liquid. A characteristic velocity $U$ can be obtained from a tangential force balance. By neglecting the thermal diffusivity and the dynamic viscosity of the gas phase, Raake et al. [22] suggested the characteristic flow velocity to be defined as

$$U = \frac{l}{\eta}\left|\frac{\partial\sigma}{\partial T}\right|\left|\frac{\partial T}{\partial z}\right|. \qquad (2.3)$$

Therefore, the Mg-number can be expressed in the following form:

$$Mg = \frac{l^2}{\alpha\,\eta}\left|\frac{\partial\sigma}{\partial T}\right|\left|\frac{\partial T}{\partial z}\right|. \qquad (2.4)$$

Interestingly, former researchers failed to agree on the definition of the mentioned characteristic length in the Re-number and the effective temperature gradient of the thermocapillary convective flow as shown in table 1. Both G. Wozniak and K. Wozniak [23], and K. Wozniak et al. [24] used $r_B$ only to represent the characteristic length in calculating the dimensionless number Mg, while Chun et al. [25] used a modified equation including both the maximum bubble radius and its vertical expansion. It should be noted that, under gravitational conditions, the injected bubble takes a non-spherical shape as it is affected by both buoyancy and hydrostatic forces from the liquid towards the gas bubble. Chun et al. [25] also used an empirical equation to determine the vertical temperature gradient near the bubble. In their configuration of a liquid matrix enclosed between two horizontal walls, where the upper one was heated and the lower one was cooled, the applied temperature gradient among the liquid matrix was calculated by an empirical equation in terms of the upper plate temperature and bottom plate temperature as described in table 1. K. Wozniak et al. [24] used a defined temperature gradient between the upper plate and a reference point $T_{ref}$, where $r_{ref}$ is the radial distance away from the center of the upper plate, and $\Delta z_{ref}$ is the vertical distance downwards the upper plate, while by G. Wozniak and K. Wozniak [23], the mean whole temperature gradient along the liquid matrix $|\partial T / \partial z|_\infty$ was simply considered.

| G. Wozniak and K. Wozniak [23] | $Mg = \dfrac{r_B{}^2}{\alpha\, \eta}\ \left|\dfrac{\partial \sigma}{\partial T}\right|\left|\dfrac{\partial T}{\partial z}\right|_\infty$ |
|---|---|
| Chun et al. [25] | $Mg = \dfrac{r_B\, z_B}{\alpha\, \eta}\ \left|\dfrac{\partial \sigma}{\partial T}\right|\left|\dfrac{\partial T}{\partial z}\right|_{ref}$<br><br>$\left|\dfrac{\partial T}{\partial z}\right|_{ref} = 0.03717(T_D\text{-}T_B),\ \Delta z_{ref} = 1$ mm, $r_{ref} = 45$ mm |
| K. Wozniak et al. [24] | $Mg = \dfrac{r_B{}^2}{\alpha\, \eta}\ \left|\dfrac{\partial \sigma}{\partial T}\right|\left|\dfrac{T_D\text{ -}T_{ref}}{\Delta z_{ref}}\right|$<br><br>$\Delta z_{ref} = 7$ mm, $r_{ref} = 45$ mm |

*Table 1: Definition of the dimensionless number Mg by different authors, where $T_D$ denotes upper plate temperature and $T_B$ denotes bottom plate temperature*

The reference temperature for the temperature-dependent fluid properties was the one measured at an adequate reference point located near the bubble periphery ($r = 8.5$ mm, $z = 2.0$ mm). This is the position where we can approximately expect a mean stable reference temperature near the bubble vicinity, where our flow phenomenon takes place. A nearer position of the reference temperature measurement to the bubble periphery could have influenced the heat transfer within the bubble vicinity taking into consideration a maximum horizontal radius of the injected air bubbles of 4.5 mm. The applied vertical temperature gradient $\left|\dfrac{\partial T}{\partial z}\right|$ is determined through temperature sensors installed inside the test cell as shown later in chapter 3.5. In the present study, the form used by Chun et al. [25] is used as both geometrical parameters of the bubble are taken into account. Moreover, the actual local temperature gradient near the bubble is used instead of the whole temperature gradient of the test fluid to determine the Mg-number. Using the same criteria used before in the available literature allows to compare our data with findings of others.

## 2.3 Oscillatory Thermocapillary Flow

Numerous authors studied thermocapillary instabilities and hydrodynamic instabilities due to the thermo-solutal Marangoni effect. These instabilities result usually at higher Marangoni numbers, where they appear normally in form of a fluctuating or oscillatory behaviour of the flow near the free surface. This oscillatory flow can be visualized through oscillations of fluid tracer particles and the temperature field around the injected air bubble.

In 1989 Raake et al. [22] used both light sheet illumination and a schlieren interferometer using a Wollaston prism to study the flow field of various viscous silicone oils around an injected air bubble under a heated wall. For a low viscosity silicone oil (AK 0.65) of Wacker® and a bubble diameter of about 8 mm, through an oscillatory movement of the interference fringes, oscillatory instabilities of the density- and the temperature fields were observed at temperature gradients of more than 0.5 K/mm. One set of oscillations of a frequency of 0.24 Hz was noticed at Mg = 12500 while, with increasing temperature gradient to more than 1 K/mm at Mg = 22000, the oscillations were transmitted to a higher frequency mode of more than 0.6 Hz. The effect of changing the size of the injected bubble was not investigated in detail and was considered to be very small compared to other parameters.

In their analysis of the Marangoni convection, Chun et al.[25] conducted a detailed experimental study of the various periodic oscillation modes of the flow of silicone oil AK 0.65 around an air bubble under a heated wall. Both shearing interferometer and flow visualization technique were used to study the flow phenomenon. Flow visualization was conducted through both horizontal and vertical light sheets to study the oscillation in both vertical and azimuthal directions. The steady flow that was noticed at $\left|\frac{\partial T}{\partial z}\right|_{z\,=\,1\,\text{mm}} \leq 0.2$ K/mm was characterized by an axisymmetric vortex, which shows up close to the air bubble interface and was followed by another secondary vortex and further vortices under the air bubble. The primary vortex was noticed to be smaller than the secondary one due to the coupling effect of both buoyancy and thermocapillary flow and confirming the findings of G. Wozniak and K. Wozniak [23]. Through the horizontal light sheet, an annular ring below the bubble was found in the center of the secondary vortex. For steady-state flow, this circular ring deformed periodically in the azimuthal direction with time. The periodical deformation of the ring and the period of the travelling wave were also discussed. Therefore, different oscillation modes of patterns of the periodical deformation were defined depending on the wave numbers.

Although the effect of the injected bubble diameter on the oscillatory flow was considered to be small [22], Chun et al. [25] draw their attention to the importance of a geometric shape parameter or bubble aspect ratio ($r_B/z_B$) to define the different oscillation modes of the flow till reaching non-periodic oscillation modes. The geometric shape parameter is defined as the ratio between the radius of the injected bubble ($r_B$) to its height or vertical extension ($z_B$). They pointed out that the critical Mg-number was different for every bubble aspect ratio as shown in figure 2.1. The oscillations in the vertical directions were found to be smaller than the ones of the azimuthal one. The mentioned oscillations were attributed to the thermal gradient driving force and dependently thermocapillary convection in the flow field around the bubble. Therefore, the temperature distribution in the bubble surrounding region was also expected to oscillate with the wave period $t_p$. Measuring temperature oscillations around the bubble using thermocouples was also recommended by the authors to get a more accurate wave period $t_p$.

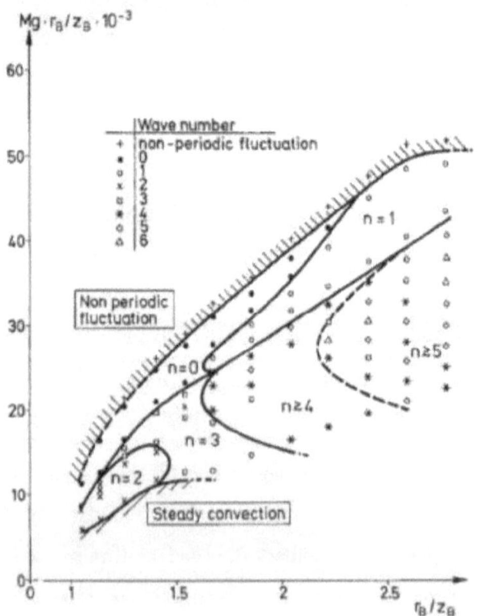


Figure 2.1: Mode selection on the plane of the modified Mg vs. the bubble geometry parameter $r_B/z_B$ [25]

Later in 2000 in a comparable study, Reynard et al. [26] studied the various oscillation modes in a silicone oil layer subjected to a vertical temperature gradient. The thermocapillary convection was analyzed using an optical measurement technique based on shadowgraphy. The variation of refractive index of silicone oil with temperature forms a line around the air bubble that represents the boundary of the thermocapillary vortex. The appearance of an oscillatory mode was determined by following the distortion of the line around the air bubble versus time during observing the flow in a vertical section. Three different oscillation modes were classified according to the complexity of the oscillation whether it is in horizontal, vertical direction, or both directions. Therefore, two frequencies of the oscillations were represented and found to depend on both the bubble aspect ratio and the Marangoni number. The influence of both temperature gradient and bubble aspect ratio on the oscillating flow behaviour and corresponding frequencies was also discussed. The value of the critical Marangoni number, where the oscillatory flow starts, was found to increase with increasing bubble aspect ratio. On the same side, later on, Reynard et al. [27] found that the critical temperature gradient, required to drive the flow into the oscillatory state, increases at low bubble aspect ratios.

The disagreement between the results of Reynard et al. [26] and that of Chun et al. [25] was attributed to differences in $v$ and thus Pr of the tested silicone oils. It was also due to the difference of the silicone oil layer thickness in both experiments. At any Marangoni number, the frequency was considered to increase with decreasing aspect ratio.

Later, using the same measuring technique, Reynard et al. [27] observed a flow oscillations frequency shift for different bubble aspect ratios at high Mg-numbers. These frequency shifts correspond to a change in the wave number, which was observed on a control screen.

In 2001, Betz and Straub [28] represented a detailed study of the thermocapillary instabilities arising in a conventional arrangement of a stationary bubble injected in various alcoholic liquids ($7 \leq Pr \leq 120$) under high vertical temperature gradients. The liquid included inside a test cell of 26 x 26 mm$^2$ cross-sectional area and 7.4 mm height was subjected to an upward temperature gradient using Peltier-modules attached to the upper and bottom copper plates. The utilization of four double-glazing glass walls allowed the direct observation of the flow field around the bubble using Particle image velocimetry (PIV). For the PIV setup, a light sheet of 1 mm thickness, generated with a He-Ne laser, was adjusted to the height of the liquid layer to illuminate the fluid around the injected bubble. Hollow glass spheres with a  mean diameter of around 100 μm were used as tracer particles. For a quantitative study of the frequency of the oscillatory flow transient temperature variations at a discrete point in the liquid near the heating plate, the temperatures were measured by a thermistor with a head diameter of 0.35 mm over a period of 23.85 s.

At low Marangoni numbers, the steady thermocapillary-buoyancy convection flow pattern consists of a clockwise primary recirculation zone near the injected bubble and another opposite direction secondary recirculation zone near the bottom part of the liquid. These observed flow vortices were considered to arise due to viscous shear stresses in the liquid. On the same side at relatively higher Mg-numbers, a shift of the primary recirculation zone towards the warm plate and a vertical movement of the secondary recirculation were also noticed.

At Mg-numbers in a range of 30000 to 35000, periodic oscillatory wavy flow patterns were noticed, and characterized by the split of the primary recirculation zone in multiple separated recirculation zones. The authors referred the reason of those oscillations to thermocapillarity that reduces the driving temperature gradient, then heat by conduction enhancing the thermocapillary convection again leading to periodic oscillatory flow. The oscillations changed from a periodic to a non-periodic manner by increasing Mg-numbers. That was also noticed by transient temperature predictions at the discrete point, where the frequency of temperature oscillations increased while increasing temperature gradients and accordingly Mg-numbers.

In 2007, Schade et al. [29] used PIV and a differential interferometer to study the oscillatory behavior of the flow field around a bubble under thermocapillary and buoyancy forces for both methanol and silicone oil AK 0.65. The experimental arrangement with liquid included between an upper heated plate and a lower one cooled, was simply used to generate an upward vertical temperature gradient. The velocity distribution around the bubble was determined applying the PIV evaluation software DaVis of LaVision® from successive video images, while the interferometer used was described before in [30]. For both liquids, there were limitations for the maximum upper plate temperature due to the vaporization temperatures of the liquids at atmospheric pressure. Consequently, the formation of small bubbles around the injected one disturbed both the flow visualization and temperature field evaluation using the interferometer. For methanol, oscillatory flow was noticed at Marangoni numbers of more than 19000 for bubble radii of 3-4 mm. The maximum temperature difference reached for the used test cell and methanol as test liquid was 66 °C, where oscillations and wavy flow were noticed around the injected bubble. On the same side oscillatory flow was noticed in case of using silicone oil AK 0.65 as a test liquid using various successive interferometer images.

The oscillations of the liquid flow around an injected air bubble were usually considered to increase when increasing the applied temperature gradient along the bubble free surface and therefore, higher Mg-numbers. However, in 2009, Schade et al. [31] observed an unpredicted restabilization of the flow around the injected air bubble at Mg = 63000 after reaching high irregular oscillations at a relatively higher Mg-number (Mg = 58000).

As mentioned by Chun et al. [25], these unexpected observations of Schade et al. [31] rely basically on the difference of the aspect ratios of the injected bubble in every case. The value of the critical Marangoni number was considered to depend on the aspect ratio of the injected air bubble as shown before in figure 2.1. Therefore, in our study the diameter of the injected bubbles should be well-controlled in order to gain correct conclusions on the behavior of the flow at higher Marangoni numbers beyond the critical one.

## 2.4  Experiments under Microgravity

Several studies have been attempted to investigate the thermocapillary convective flow under reduced gravity conditions. The motivation of all those experiments was to study the surface tension driven flow field around a free surface without the disturbing buoyancy influence of earth-bound investigations. The experiments can be classified into the following: experimental studies of the thermocapillary convection around a single or multiple stationary air bubbles, thermocapillary migration of deformable or nondeformable bubbles or droplets, and experimental microgravity studies related to thermocapillary instabilities.

To mention only an example regarding bubbles or droplets movement under the thermocapillary force, G. Wozniak [32] studied droplet migration and movement because of Marangoni convection inside a liquid cavity under a vertical temperature gradient in reduced gravity conditions. In brief, two identical test cells were placed over each other to permit doing two different tests at the same time. They were of $60 \times 26$ mm$^2$ cross sectional area and 15 mm height. The test cells were filled with the continuous phase liquid aqueous ethanol. Five different sized Paraffin liquid droplets were injected in each test cell. The sounding rocket, on which the experiment was performed, provided approximately 7 minutes of reduced gravity. Therefore, a stable temperature stratification (i.e., heating the fluid from the top and cooling it from the bottom) was achieved after about 3 min of reduced gravity, only by preheating the test cell on the ground before launching. The stable linear temperature profile was proved using different thermocouples placed inside the test cells at different heights.

Particle speeds were determined by tracking each drop over a distance of a few millimetres and then calculating the slope of the plot of position versus time using the method of least squares. As predicted, the drops were attracted towards the heated plate and were found to be spherical because of the reduced gravity experimental condition. In addition, the droplet moving velocity increased at larger droplet sizes. However, in comparison with the theoretical model of Young et al. [33], the experimental droplet speed values were relatively low. The author regarded this difference to the neglected surface dilatational viscosity and the shear viscosity in the theoretical formulation of the problem.

In 1996, G. Wozniak et al. [34] studied flow characteristics of surface tension driven flow around a bubble under low gravity conditions. The experiment was conducted during the TEXUS 33 sounding rocket flight, which offered a microgravity period of approximately 6 min. An air bubble of 4.5 mm diameter was injected in a liquid matrix of silicone oil AK 10 of square section of 50 x 50 mm$^2$ and 30 mm height. Before bubble injection, a stable vertical temperature gradient of 4.8 k/cm was established by heating the upper copper plate and cooling the lower one. Temperature gradients were measured using various thermocouples, which were vertically installed inside the test cell at different heights. Flow visualization using liquid crystals as tracer particles provided both velocity and temperature fields simultaneously [35]. The injected bubble shape was quite spherical due to the absence of the hydrostatic pressure resulting from gravity, which flattens the shape of the bubble in the vertical direction in the earth-bound investigations. The jet-like flow was formed by the action of the thermocapillary force (the only force driving the flow). In comparison to the same experiment done in normal gravity, the isotherms were far beyond the bubble surface [34].

Later in 1999, in addition to their experimental work to study the thermocapillary flow instabilities under normal gravity conditions, Reynard et al. [27] were the first to observe flow oscillations under reduced gravity conditions. They observed periodic oscillations around an air bubble immersed in a low-Prandtl-number silicone oil layer under a heated wall. Although the initiation of oscillatory flow was found to be independent of gravity and buoyancy, the gravity level was found to modify the oscillation. Unlike the flow oscillations under normal gravity, the oscillations were found to be asymmetrical with odd wave numbers under reduced gravity. The frequency of oscillation depended on the gravity level. The period of thermal oscillation decreased while increasing the gravity level. Its frequency was found to increase with the gravity level. Comparing the results under reduced gravity with those obtained on ground showed that the bubble aspect ratio was not a relevant parameter when the gravity level varied.

# 3  EXPERIMENTAL SETUP AND METHODOLOGY

## 3.1  Preface

Flow visualization techniques are considered to be effective methods for measuring fluid flow velocity fields and capturing flow streamlines. Researchers have developed both visualization tools and image evaluation or processing techniques to enable accurate velocity measurements and proper utilization in different experimental setups despite their limitations. Compared to other techniques for velocity measurements as hot-wire anemometry (HWA) and laser-Doppler velocimetry (LDV), PIV has shown its capability in modern experimental fluid mechanics as discussed by Adrian [36]. He showed that PIV provides spatial derivatives, flow visualization, and capability for the spatial correlation offered, which are not provided by other techniques. That annual review article considered many different issues including the best optical flow analysis, particle tracking and spatial resolution to study complex flows such as turbulence. Moreover, Keane and Adrian [37] presented some guidelines to other important parameters such as the tracer particles optimal concentration and exposure time delay between the two light pulses with respect to the size of the interrogation domain, and the light-sheet thickness to provide an optimal measurement technique. The topic of measurements uncertainty was discussed by Adrian [38] who defined the performance of a given PIV system using an uncertainty principle, which states that a PIV system can achieve high spatial resolution only with the penalty of reduced relative measurement accuracy, and vice versa.

The implementation of tracer particles with light sheet illumination of a certain plane of the flow field was for example used by Rösgen et al. [39], who presented an algorithm for evaluating the flow field images based on the 2-D fast Fourier transform for fringe analysis.

This technique was later used by K. Wozniak and G. Wozniak [40]; and G. Wozniak and K. Wozniak [41] to measure flow field velocities in a Bénard-convection and a thermocapillary-buoyancy convection, respectively. The experimental results of [40] agreed well with the theoretical ones of Oertel [42]. Moreover; G. Wozniak et al. [35] applied a novel combined measuring technique to measure both velocity and temperature fields of thermocapillary-buoyancy convection using liquid crystals as tracer particles. According to the fact, that the wavelength and thus the color of reflected light by liquid crystals depends on the crystal temperature, a quantitative temperature evaluation technique was implemented to detect the isotherms of the flow field.

---

Regarding temperature measurement techniques, many authors used interferometry in thermo-convective experiments [43]–[45]. G. Wozniak [46], for example, measured the temperature distribution after injection of a bubble under an applied upward vertical temperature gradient on the liquid matrix using a shearing interferometry. He proved that the thermocapillary force (surface tension force) drives the flow as temperature isotherms accumulated near the bubble surface.

The goal of this work is to study the instabilities of the convective thermocapillary flow around an injected bubble under a heated wall under relatively high temperature gradients. Therefore, an accurate description of the flow behaviour including both velocity and temperature distributions around the bubble is indeed required. The shape of the thermocapillary flow vortices near the bubble and the transient temperature variations at different locations near the upper plate can provide information about the stability of the flow. Accordingly, induced flows can be categorized into laminar, periodic oscillatory, and non-periodic oscillatory depending on different conditions as the driving applied vertical temperature gradient and the injected bubble dimension (Mg-number).

In the following chapter, the experimental setup used to pursue the experiments as well as different techniques used for flow visualization, temperature, and velocity measurements are presented and discussed.

## 3.2  Test Cell

In order to experimentally investigate the turbulent thermocapillary flow around a bubble under a heated wall using optical measurement techniques, a certain adjustment of the test cell was being used to pursue that experimental study as shown in figure 3.1. The applied test cell is similar to that described by Schade et al. [29].

It is, in brief, a rectangular cavity of 50 x 80 mm horizontal dimensions ending with two horizontal copper plates of 14 mm thickness. The side walls were cut from a polycarbonate block by means of water jet cutting and all sides were finely polished until optical accessibility was achieved, while the interior walls had to be completely polished by hand. The use of a single-block glass wall allowed us to carry out our experiments at higher pressures, avoiding any sealing problems caused by adhesives between the sidewalls. With the help of 4 corner screws, the upper and lower copper plates are tightly sealed to the glass walls by means of gaskets inserted between them. In order to reach higher temperature gradients in the test cell, the clear height of the liquid matrix was adjusted to 15 mm by introducing further copper plates on the lower copper plate. For injecting air bubbles, there is a 1 mm diameter hole in the center of the upper copper plate connected to a flexible plastic hose. The top plate contains two different through-tubes for pressure equalization.

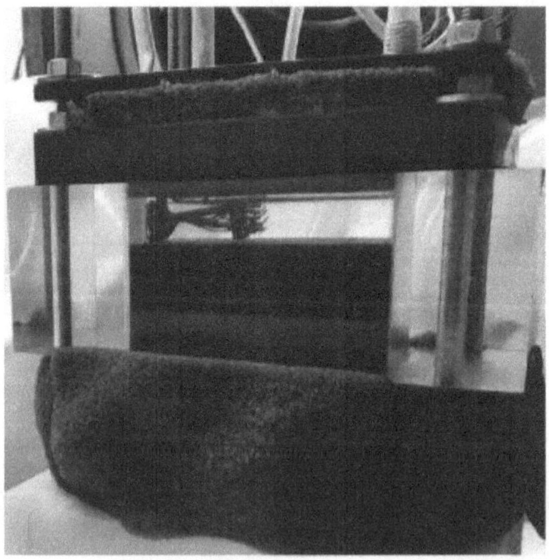

*Figure 3.1: Photo of test cell structure*

The upper and lower boundaries were maintained at two different uniform temperatures to generate a vertical temperature gradient, which is required for generating the thermocapillary flow. The lower plate was cooled and maintained at a lower temperature ($T_c$) by means of a LAUDA compact low-temperature thermostat, of type RK 8 CS. Using an ethanol-water mixture and a maximum pump pressure of 0.3 bar, the operating temperature could reach about -15 °C depending on the ambient temperature and other parameters like the heat flux applied by the upper copper plate heating resistor. On the other hand, the upper plate was maintained at a higher temperature ($T_w$) using an electrical resistance heater. The temperature of the top plate was measured using another thermocouple attached to a middle hole of the top plate. The temperature of the upper plate could also be adjusted by means of a small control unit which receives the temperature values and controls the intensity of the electrical signal applied to the electrical resistance.

### 3.3  Pressure Chamber

#### 3.3.1 Necessity of Gauge Pressure Experiments

In general, observing thermocapillary instabilities in the flow behaviour around a bubble injected under a heated wall requires reaching higher Marangoni numbers. Accordingly, authors tried to use different low viscous liquids like alcohols and silicone oils under relatively high applied upward temperature gradients. These high temperature gradients could be either reached using test cells of small heights or by increasing the upper wall temperature to the maximum and decreasing the lower wall temperature to the minimum available one. A key problem regarding small height test cells to achieve higher temperature gradients is that the bottom wall effects the behaviour of the flow. The shortcomings of those methods are clearly recognized and discussed in the current study (see chapter 4.1.3). Therefore, in our experiments we used relatively intermediate heights of the test cell with a maximum reachable temperature difference between the upper and lower copper plates to achieve high Mg-numbers.

The volatility of the low viscosity fluid used in the experiments, described before, plays an obvious rule and can be easily noticed. As described in past studies, it was noticed that at high temperatures the injected bubble began to flatten out then forming another small bubble near to the injected one which disrupted the aimed examined flow field around the bubble as shown in figure 3.2. This accordingly disturbed the flow field to be investigated and both, PIV and shadowgraph images, respectively. Therefore, increasing the temperature of the heated plate is limited to vapour pressures of the examined liquids at high temperatures.

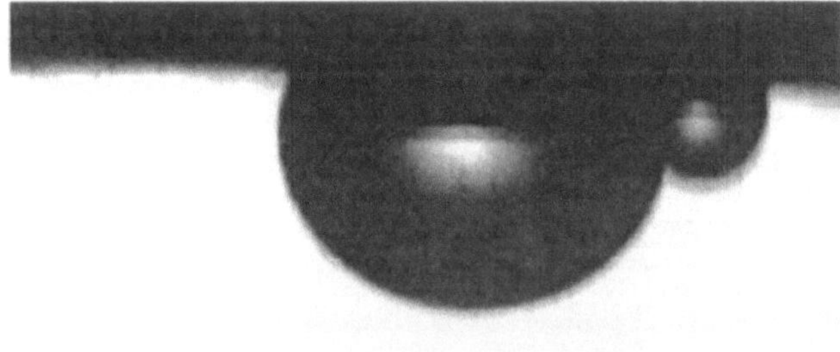

*Figure 3.2: Vaporization effect on bubble formation* [47]

As mentioned before, Schade et al. [29] reached a maximum temperature of 53 °C and 45 °C of the upper heated plate while using alcohol and silicone oil AK 0.65, respectively, as a test fluid. They also recommended to pursue the experiments under higher pressure just to avoid any formation of disruptive bubbles. According to vapour pressure-temperature curves of different fluids shown in figure 3.3, it is noticed that for silicone oil AK 0.65 a gauge pressure of $\Delta p = 1$ bar is enough to shift the vaporization temperature to the range near 100 °C.

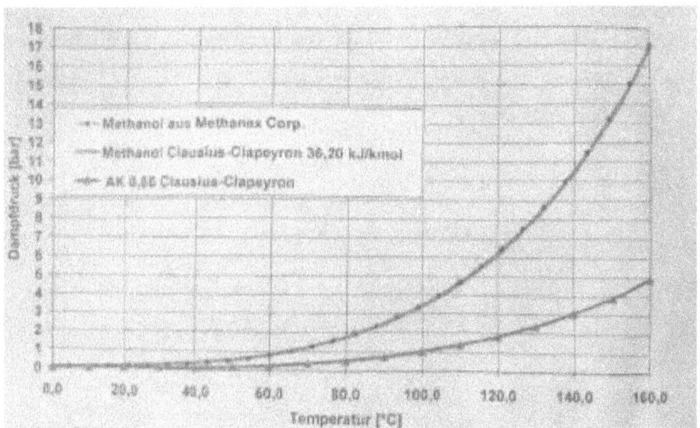


*Figure 3.3: Vapour pressure - temperature curves of different volatile liquids* [47]

According to the various trials of Schade et al. [31] to establish a classic high-pressure test cell using a supporting frame from Polycarbonate or wood, it was unpractical to design a fully sealed high-pressure test cell. The most successful trial was to establish a completely sealed test cell, but unfortunately, the test cell could not preserve the high pressure of 4 bar although no cracks were noticed. Consequently, it was suggested to use a pressure chamber that can mechanically preserve pressurized air. Moreover, through the continuous supplying of pressurized air inside the pressure chamber a stable high pressure could be established regardless the chamber is fully sealed or not. The pressurized air around the test cell can ensure also the sealing of our test cell, and no liquid can escape while injecting the air bubble under relatively higher surrounding air pressure.

### 3.3.2 Pressure Chamber Description

The pressure chamber, shown in figure 3.4, is in brief a cubical cell of internal side length of 37 cm, with four mountable steel lids. The upper boundaries and sidewalls consist of different sealed connection parts to access different utilities for heating and cooling the test cell plates, air valves control, pressurized air supply, bubble injection, and temperature and pressure measurements. The pressure chamber is adequate to PIV experiments as two glass windows of optical quality are mounted on the right-side wall and the front sidewall for light accessibility of the laser beam and the PIV camera. All connection parts are completely sealed to permit a stable pressure inside the pressure chamber. The pressure chamber is mechanically designed to maintain a maximum gauge pressure of about $\Delta p = 4$ bar, without any risks of damage or air leakage (see the appendix).

*Figure 3.4: Pressure chamber*

### 3.3.3 Working Principle

Pursuing experiments inside the mentioned pressure chamber allows reaching high temperature gradients; however, the stability of the injected air bubble is obviously affected. The continuous supply of air towards the pressure chamber to ensure stable air pressure influences badly the interface between the silicone oil and air. The pressurized air inside the chamber seeks usually to be saturated with silicone oil vapour and vice versa. Therefore, the two pressure-equalization through-tubes should be closed after reaching the chamber pressure. This was achieved using an electric air valve (B) connected to the tubes as shown in figure 3.5. Another electric air valve (A) was used through the bubble injection air supply line in a position near the test cell. That ensures a constant pressure inside the bubble injection line regardless of any leakage, arising from the air supply line or any unsealed connectors. That also ensures a stable and constant bubble size. The bubble injection line came from the outside of the pressure chamber with an initial pressure equal to the chamber pressure. In addition to that an injection syringe was attached from outside to control the size of the injected gas bubble.

To carry out the experiments the test cell was filled with the transparent matrix liquid silicone oil AK 0.65 ($v = 0.65. \ 10^{-6} \ m^2/s$ at T = 25°C, Pr = 7.16) as test liquid. Data for the values of $\rho$, $\alpha$, $v$, $\eta$ and $\sigma$ were provided by Wacker silicones®. Then, the maximum permissible temperature gradient under atmospheric pressure was established. After that, valve (A) was closed and valve (B) was opened and the air pressure inside the chamber began to increase till reaching the wanted pressure ($\Delta p$). Later, valve (B) was closed and the temperature of the upper plate was increased again until reaching the wanted temperature ($T_w$). When reaching a stable steady temperature gradient inside the test cell, a gas bubble of controlled size was injected using the injection syringe from outside by opening the air valve (A) and closing it after injection.

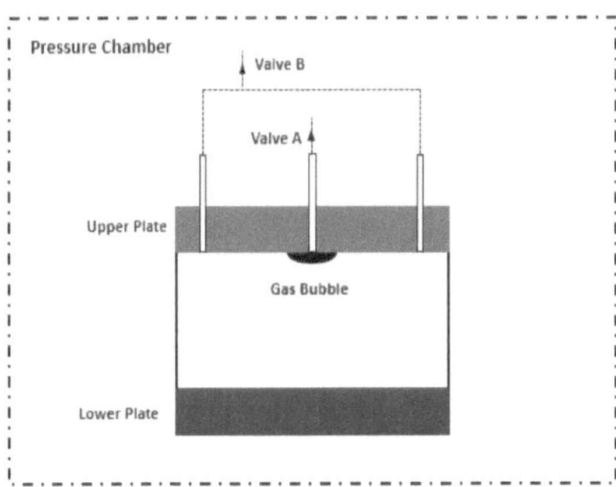


*Figure 3.5: Schematic of the experiment procedures inside the pressure chamber*

### 3.3.4 Experimental Impediments Inside the Pressure Chamber

By increasing the pressure inside the pressure chamber and accordingly the temperature of the upper copper plate, an invisible layer was noticed as shown in figure 3.6. This invisible layer was observed through shadowgraphy, where the full bubble contour at the upper partition near the upper plate was invisible. One can refer this to the refractive index difference between pressurized air where the test cell was placed and atmospheric-pressure air where the camera was placed. The thickness of the mentioned invisible layer ($L_b$) increased usually by increasing the pressure inside the pressure chamber as shown in figure 3.7. It was, accordingly, recommended to tilt the recording camera vertically upwards towards the test cell, so that good images can be taken.

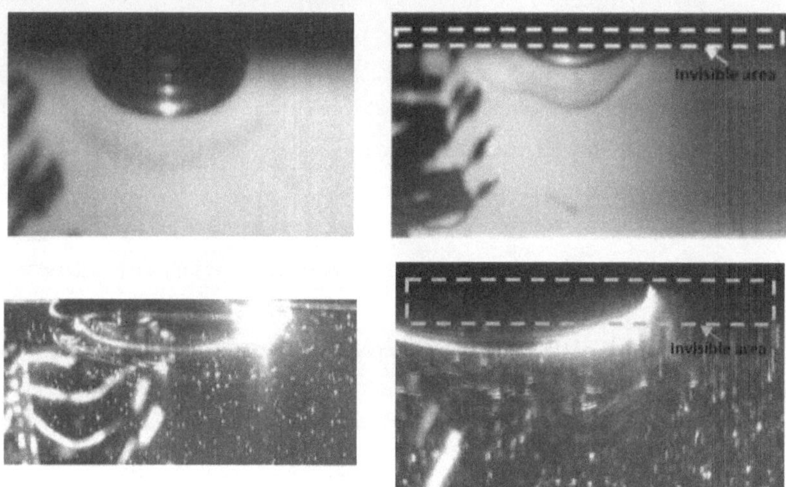


*Figure 3.6: Formation of blind layer at higher pressures (right side) comparing to relatively good images (left side) of both PIV and shadowgraph images*

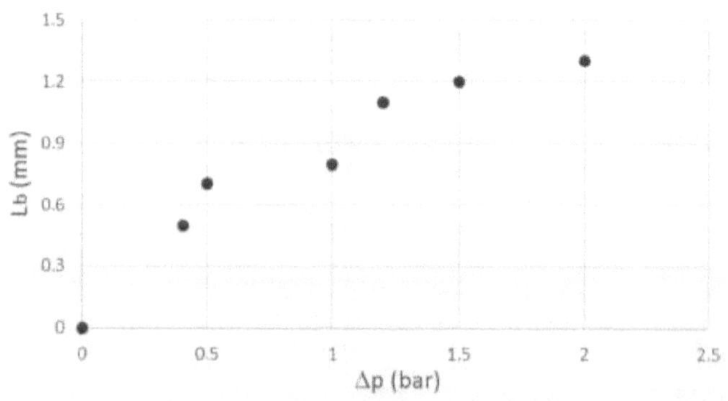


*Figure 3.7: Thickness of the blind layer versus the pressure of air inside the pressure chamber*

## 3.4 Particle Image Velocimetry (PIV)

In general, this method depends on introducing seeding particles to the fluid where these small particles move with the local fluid velocity of the flow. These particularly small seeded particles have small body forces compared to the surface forces. Accordingly, the seeding particles, particularly with very small diameters, have no response to buoyancy forces and are considered neutrally buoyant with respect to the fluid medium. By illuminating a thin light sheet across the flow stream, the position of the seeded particles inside the flow field can be captured using a perpendicularly positioned camera to the plane of illumination. This technique basically evaluates the particle displacement $\Delta s$ over a time period $\Delta t_{1\text{-}2}$ forming velocity components in the plane of illumination. Since particle sizes are very small, there would be several particles in a small interrogation area selected by the evaluating software for velocity determination, where they are indistinguishable. The measured displacement is a statistical quantity that is applicable to the particle collection unit. Thus, the obtained local velocity is a group velocity of these small particles. The most common used statistical methods are based on cross-correlation between a pair of images separated by the time interval $\Delta t$. To get a more accurate velocity estimation, one can use shorter time intervals between the images. For that, the illumination using conventional light sources is not practical, and it is recommended to use pulsed lasers.

The fact that PIV is a non-intrusive velocity measuring technique depends on the neglectable influence of the tracer particles on the flow behaviour. It is also necessary to ensure that flow properties are not influenced when implementing the seeding particles. A proper seeding should ensure a spatially uniform homogeneous distribution of tracer particles within the fluid. The density and size of the tracer particles should be adequate to give high quality flow field images. The applicable diameters of the seeded particles vary according to the used fluid in every experiment to follow the original local fluid velocity. To eliminate any velocity lag, the particle density should ideally match that of the fluid specially by micron-sized particles, where the surface forces are dominant. Under gravity, the particle weight can cause errors in the velocity measurement. To ensure that the particles would follow the main flow without excessive slip, the calculated settling velocity $U_\infty$ of the tracer particles should be small compared to the expected flow velocities. Assuming that Stokes law of drag is applicable, the settling velocity is given by

$$U_\infty = \frac{g\, d_p^{\,2}(\rho_p - \rho_f)}{18\, \eta_f} \,, \qquad\qquad\qquad (3.1)$$

where $\eta_f$ and $\rho_f$ denote the fluid viscosity and density, respectively, and $d_p$ and $\rho_p$ denote the tracer particle diameter and density, respectively. For the present set of experiments, where Borosilicate glass particles of mean diameter of 10 µm and density of 1.1 g/l were used as tracer particles within a liquid matrix of silicone oil AK 0.65, the settling velocity was estimated to be less than 10 % of the expected flow velocities of 0.1 mm/s in the case of steady laminar thermocapillary convection and less than 1 % of the expected flow velocities of 0.1 cm/s in the case of non-periodic thermocapillary convection.

In the present experiments, PIV measurements were carried out at a selected plane perpendicular to the upper and lower walls of the test cell as shown in figure 3.8. The test cell was illuminated using a vertical laser light sheet of 3 mm thickness originating from a low power laser source of 300 mW passing through the mid-plane of the test cell where the bubble was injected from the upper copper plate. However, a smaller light sheet thickness was also recommended to minimize the effect of the out-of-plane velocity component. The illuminated light sheet was shifted slightly behind the bubble tip surface to avoid any disturbing light reflections at the lower periphery of the bubble. A CCD camera was positioned perpendicular to the plane of illumination.

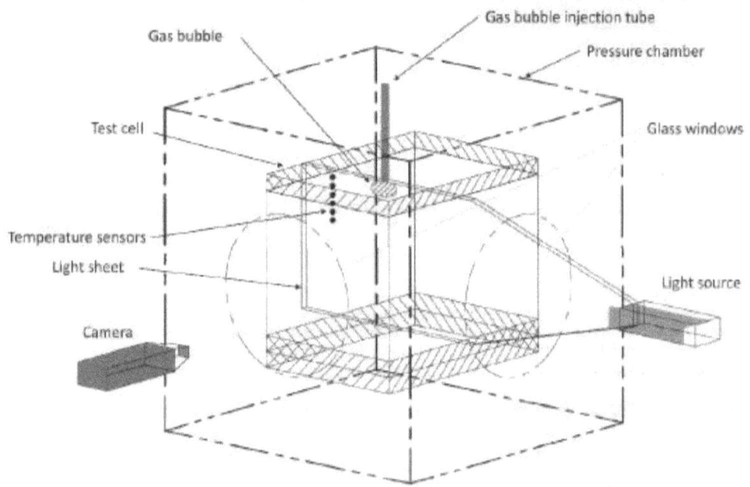



Figure 3.8: Schematic of experimental PIV setup using the pressure chamber

The assembly of the 12-bit Basler camera and frame grabber with a frame speed of 180 fps was used for acquisition of the PIV images. The used camera has a resolution of 2048 x 2048 pixels. In case of pursuing experiments inside the pressure chamber, a long-distance microscope of LAVISION® was attached to the camera to provide adequate magnification and image resolution (1 mm = 316 pixel) within the available long working distance. The field of view of these images only showed the fluid flow and corresponding velocity fields around half of the bubble periphery. Otherwise, a lens of 80 mm focal length was attached to the camera and a relatively larger field of view of the flow field around the bubble was provided, where the magnification of particle image depends on the position of the imaging lens. For accurate local velocity measurements, a shorter capturing time of the order of microseconds for each frame was applied. However, for capturing tracer particles trajectories showing the flow field behaviour, a longer capturing time for single frames was selected.

Regarding the processing of images, the successive camera images were evaluated using the Davis 8 software provided by LAVISION® to generate velocity vector field images. This software divides each recorded image into small interrogation areas. After that, by cross-correlation between two corresponding interrogation areas in subsequent images, a correlation map of tracer particle displacements is generated to create a velocity vector flow field image. In general, data generated from PIV carries superimposed noise during recording of PIV images. The mentioned noise can be due to different reasons like non-homogeneity of the light sheet or non-spherical particles. The validity of the final velocity distribution depends also on data processing parameters like the selected interrogation area and the image interpolation and peak deformation.

In general, PIV measurements were performed at moderate temperature gradients (moderate Marangoni numbers) where the refractive index change is low and has no have effect on the velocity measurements.

## 3.5  Temperature Measurements

The thermocapillary flow in our configuration depends on the temperature gradient subjected to the bubble interface and thus the surface tension gradient. Therefore, accurate and adequate temperature measurements are required. In the current study, measuring the temperature at different locations inside the liquid matrix helped us to determine the applied undisturbed temperature gradient near the bubble. Moreover, measuring the transient development of the temperature at discrete points around the bubble, where the thermocapillary forces are dominant, gave us a good image about the flow field behaviour as shown later in chapter 4.1.1. In past studies, some authors used optical methods to measure the temperature distribution within the test cell using liquid crystals [23], [24], [35], [41] or interferometry [30], [43]–[45]. Using liquid crystals for measuring temperatures is basically limited by the fact that suspended crystals and the liquid matrix should have similar refractive indices. For an Interferometer, the preliminary investigations indicated that the polycarbonate glass, of which the test cell used in our experiments was manufactured, has such high internal material tensions and inhomogeneity that made this measuring technique not applicable in the current configuration as shown in figure 3.9 and mentioned before, [31].

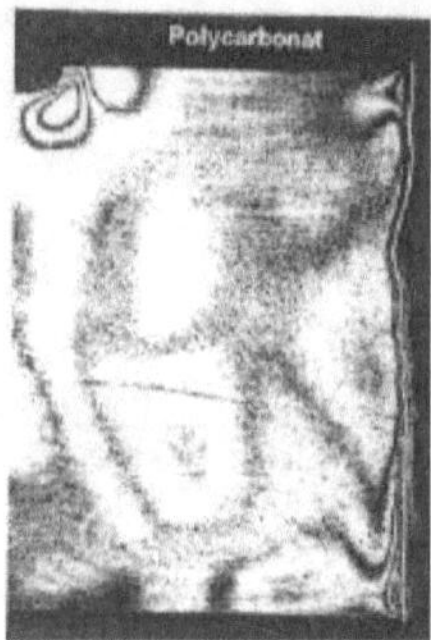


*Figure 3.9: Additional interference fringes while using Polycarbonate glass* [31]

### 3.5.1 Temperature Sensors

In the current configuration, where temperature values at a single point deviate rapidly with time, using temperature sensors with good accuracy and shorter response time was the best choice to trace that fluctuating temperatures. Basically, resistance temperature detectors (RTDs) are temperature sensors that contain resistors which change their resistance value as the temperature changes. Accordingly, they have been used for many years in laboratory and industrial processes and have developed a reputation for accuracy, repeatability, and stability. On the other side, thermocouples are temperature sensors of a two-wire junction that creates a voltage as its temperature changes. They are characterized by their short response time due to the small volume of the junction and their lower accuracy compared to RTDs. In our study, temperature measurements of both RTDs and thermocouples at a moderate Marangoni-number showed good consistency as shown in figure 3.10.

In our experiments, six identical chip RTDs, Pt100 class B, which provide a measuring tolerance of $R_0$: ± 0.12 % according to DIN EN 60751, were inserted vertically at different distances from the upper plate near the mid-line to measure the undisturbed temperature gradient near the bubble. Temperature fluctuations at a discrete point B (r = 8.5 mm, z = 4 mm) near the upper plate around the bubble were measured, where r is the radial distance away from the center of the upper plate, and z is the vertical distance downwards the upper plate as shown in figure 3.11. Temperature readings were recorded every 100 ms. First, a bubble of a certain diameter was injected into the liquid matrix after reaching a stable vertical temperature gradient. Then, temperature readings were recorded after 30 seconds from the injection. That allowed examining the periodicity of temperature fluctuations of the fluid around the bubble, which was recommended by Chun et al. [25] to give a better understanding of the oscillatory state of the thermocapillary flow.

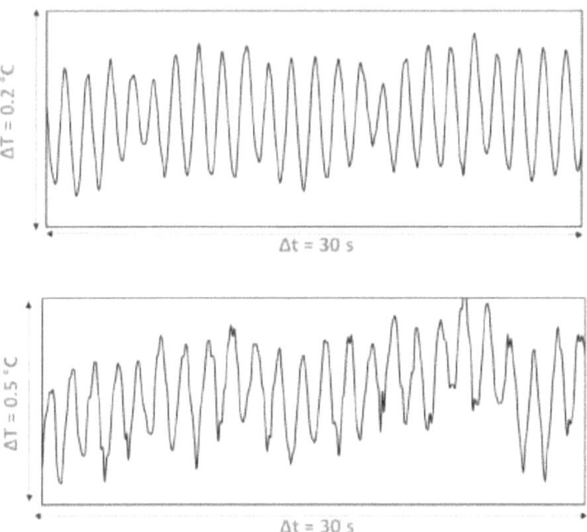



*Figure 3.10: Temperature fluctuation plots at different positions near the injected bubble (upper, using TC) & (Lower using RTD), $r_B$ = 6 mm, $z_B$ = 2.69 mm, $\frac{r_B}{z_B}$ = 2.26, $\Delta p$ = 0 bar, $\left|\frac{\partial T}{\partial z}\right|_{act}$ = 1.86 K/mm, Mg = 42200*

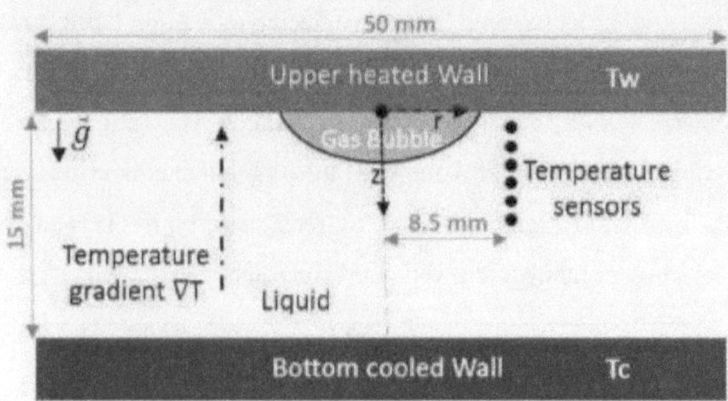


*Figure 3.11: Schematic of the location of temperature sensors inside the test cell*

### 3.5.2 Temperature Measuring System

In the current measuring setup, the RTDs were connected to a 16-bit NI 6224 Data acquisition device through temperature detector (RTD) input channel modules, named SCC-RTD 01, which were installed inside a SC-2345 signal carrier. These modules are designed to accept signals from the 4-wire RTDs, where each input channel includes a 30 Hz Butterworth low pass filter and a 1 mA excitation current source. An included instrumentation amplifier with differential inputs ensures a fixed amplification gain of 25 resulting in a maximum input voltage of 400 mVDC. The signal carrier was connected to the NI PXI-1033 chassis, which has a built-in x1 MXI-Express interface that can be cabled to a remote system with a host card using a x1 MXI-Express cable to control the NI PXI-1033 chassis as shown in figure 3.12. Through the applied LabVIEW software, the readings of the temperature measurements for the different sensors could be recorded at different time steps.

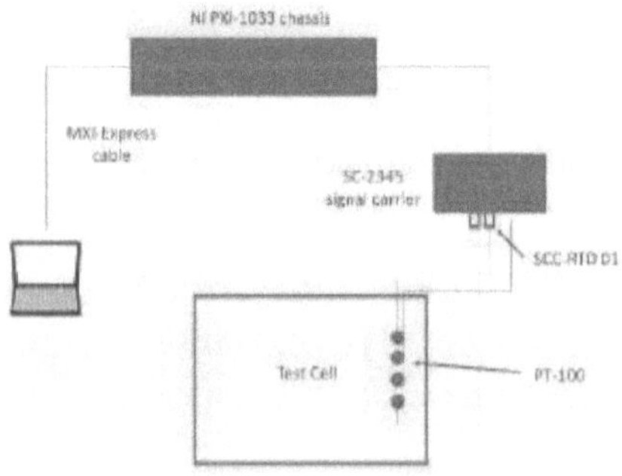


*Figure 3.12: Schematic of temperature measurement system*

## 3.6 Shadowgraphy

This measuring technique is used to give a simple image of the bubble-silicone oil system. In general, it does not need any optical component except a light source and a recording film onto which the shadow of the captured field is projected. The refractively deflected light ray appears brighter than the dark undeflected one on the camera sensor of spatial resolution of 1 mm = 82.7 pixel. Therefore, this technique gives us the real dimensions of the injected air bubble without any distortions of light reflections which occurred when using the PIV technique. More details about this technique can be found in [48].

This method depends on subjecting a beam of light towards the test cell, where the refractively deflected light ray appears brighter than the dark undeflected one on the recording film as shown in figure 3.13. The shadow effect follows since the visible signal dependent on the second derivative of the refractive index of the fluid. The setup used in the current study for visualizing the thermocapillary vortex boundary contour around the gas bubble is similar to the setup used before by Xi et al. [49] to proof that heat plumes initiate the turbulence within a Bénard-Rayleigh convection.

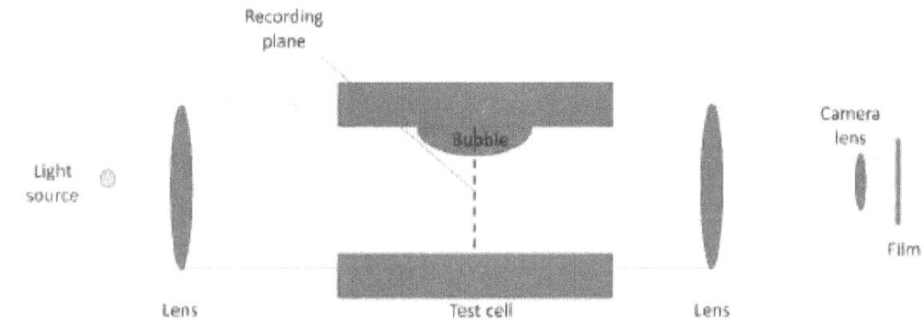


*Figure 3.13: Schematic of shadowgraphy setup*

Through the shadow pattern, one could additionally identify different oscillatory modes of the thermocapillary vortex boundary contour (see figure 3.14), which appears around the injected bubble due to differences in the refractive indices representing the borders of the convective thermocapillary recirculation zones. This boundary contour shows up due to differences in the refractive indices between the area of the disturbed temperature gradient near the bubble periphery because of the induced thermocapillary convective flow and the area of undisturbed stable temperature gradients relatively far from the flow area. The black structures, displayed left in figure 3.14, correspond to the installed temperature sensors and do not present any thermocapillary flow structure. That could be also noticed in other shadowgraph images presented in chapter 4.

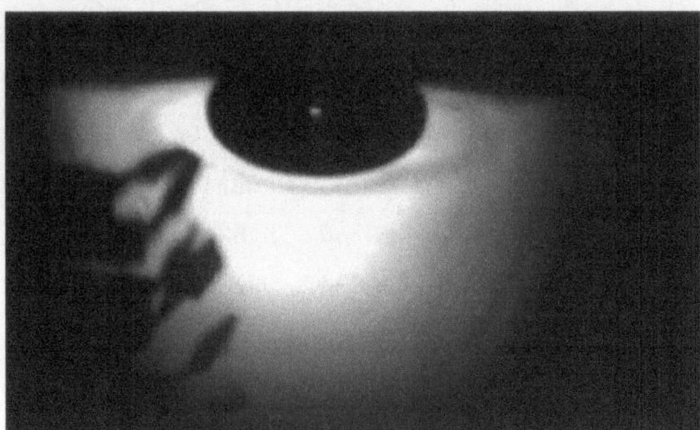

*Figure 3.14: Shadowgraph image of the thermocapillary flow around an injected bubble*

## 3.7 Horizontal Plane Visualization

This methodology aims at giving a clearer description of the periodicity of the thermocapillary convective flow. As mentioned before by Chun et al. [25], a ring-shaped concentration of tracer particles was observed. The motion of this illuminated ring can give us a clear indication about the flow periodicity (see chapter 4.1.2).

The experimental setup of this technique is the same of PIV (vertical flow visualization) except that the laser light sheet of 3 mm thickness was subjected horizontally just under the injected bubble and a 45° mirror arrangement was introduced under the bubble. Accordingly, the tracer particle motion in different planes parallel to the heated wall were observed as shown in figure 3.15.

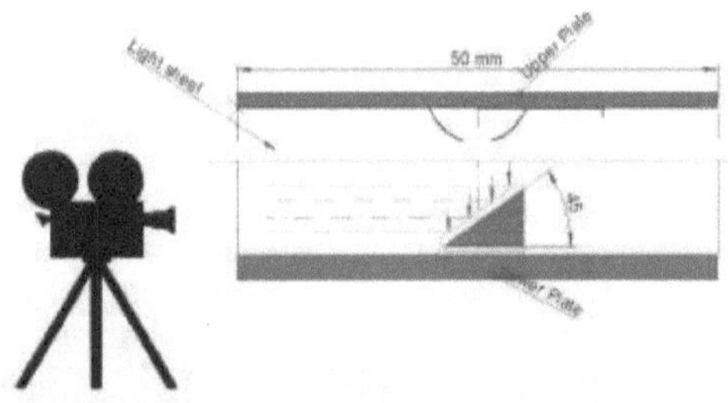


*Figure 3.15: Schematic of horizontal plane visualization arrangement*

Concurrent visualization of both vertical and horizontal planes under the bubble provided 3d tracing of the tracer particles in both vertical and horizontal cross-sections. The observed ring-shaped concentration was found to move in both vertical and horizontal planes. Figure 3.16 shows an example of the concurrent visualization of both vertical and horizontal planes of the thermocapillary flow around the injected bubble. It assures that the captured bright ring-shaped particle in the horizontal plane (lower part) corresponds to the tracer particles concentration shown beneath the bubble (upper part).

Through consecutive images of the mentioned ring-shaped concentration of tracer particles in vertical and horizontal planes, the periodic oscillatory fluid flow around the bubble could be categorized according to different wave numbers.

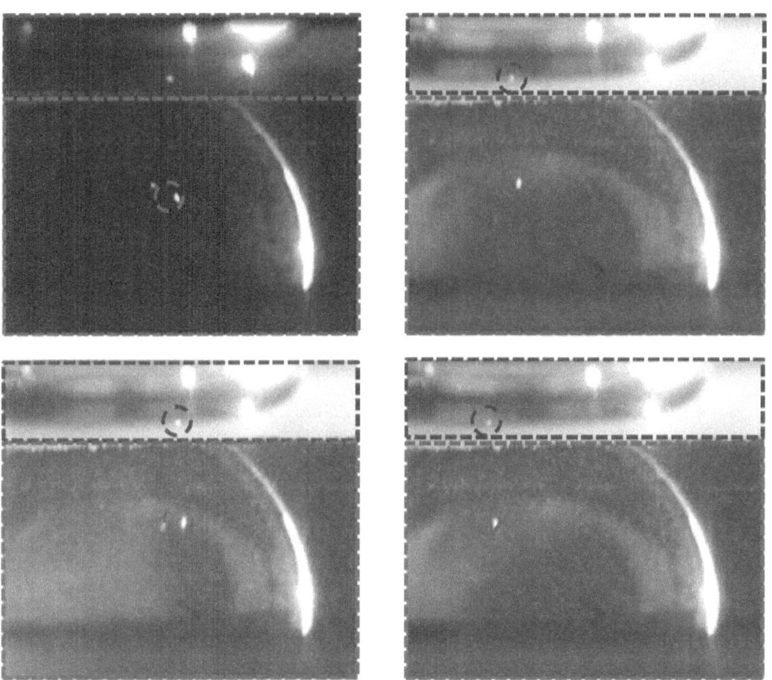

*Figure 3.16: Different images of ring-shaped concentration of tracer particles in vertical visualization (in blue) and horizontal visualization (in red)*

# 4  RESULTS AND DISCUSSION

## 4.1  Laminar and Periodic Oscillatory Thermocapillary Flow

### 4.1.1 PIV and Temperature Measurement Results

Here the results of different experiments using silicone oil AK 0.65 as a test liquid inside our test cell of dimension 50 x 80 x 15 mm³ are presented and discussed. It should also be noted that the temperature, velocity, and shadowgraph measurements were pursued during separate experiments but at the same corresponding boundary conditions.

First, it is noticed that the shape of the injected bubble is affected by both buoyancy and hydrostatic forces from the liquid towards the gas bubble. Therefore, no geometrical bubble shape similarity among different bubble radii could be achieved in our experiments as shown in table 2.

| $r_B$ (mm) | $z_B$ (mm) | Geometric shape parameter ($\frac{r_B}{z_B}$) |
|:---:|:---:|:---:|
| 2 | 1.899 | 1.053 |
| 2.5 | 2.2 | 1.136 |
| 3 | 2.4 | 1.25 |
| 3.5 | 2.5 | 1.4 |
| 4 | 2.6 | 1.538 |
| 4.5 | 2.7 | 1.667 |
| 5 | 2.7 | 1.852 |
| 5.5 | 2.7 | 2.037 |
| 6 | 2.7 | 2.222 |
| 6.5 | 2.7 | 2.407 |

*Table 2: Relation between the maximum radius of the injected air bubble $r_B$ (mm) and its vertical expansion $z_B$ (mm) for silicone oil AK 0.65*

In general, by applying an upward temperature gradient across the fluid matrix, where an air bubble is later injected after reaching the aimed stable temperature gradient, liquid particles around the bubble are peripherally attracted downwards to regions of low temperature and high surface tension, and then pushed upwards by buoyancy forces forming two symmetric primary vortices near the bubble surface as shown in figure 4.1. On the right side of the bubble the motion of the particles in the vortex is anticlockwise, and vice versa on the left side. These primary flow vortices enhance the formation of other lateral secondary vortices beside the primary ones. The pattern and shape of the flow around the bubble are obviously affected by the applied temperature gradient and the geometric shape of the injected bubbles. At low applied temperature gradients, the flow pattern is symmetrical around the vertical bubble mid-plane. However, the flow around the bubble seems to be rotational and is characterized by its laminar behaviour at relatively low temperature gradients. In addition to that, this steady thermocapillary-buoyancy flow is characterized by a stable temperature distribution around the bubble. That can be noticed also through the temperature plots resulting from temperature measurements through the build-in temperature sensors, which were placed inside the fluid domain near the bubble, shown in figure 4.2, where the temperature at point B, located beneath the upper plate ($r = 8.5$ mm, $z = 2$ mm) near the injected bubble, remains approximately constant. At $Mg = 10320$, the temperature variation at point B is considered to be constant as it deviates in a temperature range of 0.02 K.

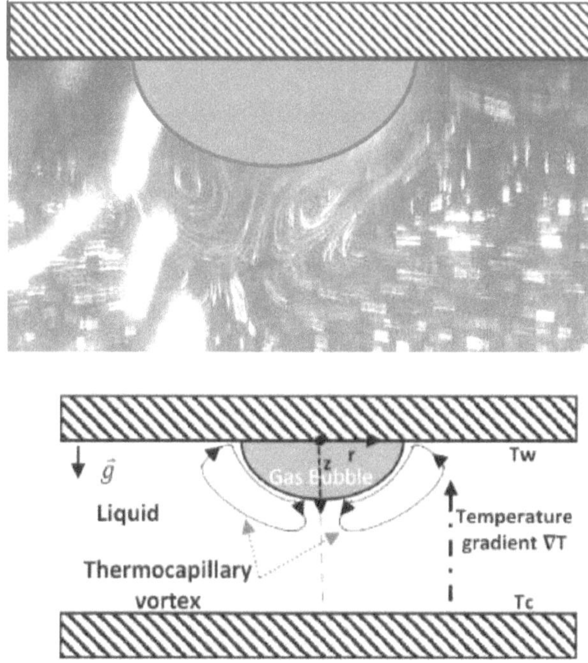



*Figure 4.1: Schematic of thermocapillary bubble convection under a heated wall*

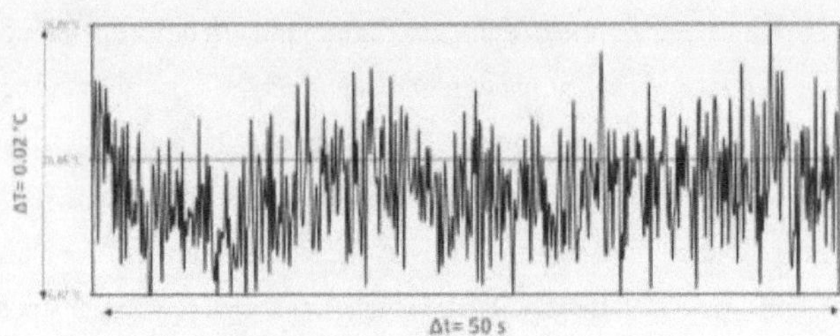


*Figure 4.2: Temperature fluctuation plot at point B in the liquid, $r_B$ = 8 mm, $z_B$ = 2.75 mm, $\frac{r_B}{z_B}$ = 2.91, $\Delta p$ = 0 bar, $\left|\frac{\partial T}{\partial z}\right|$ = 0.32 K/mm, and Mg = 10320*

At relatively higher temperature gradients, fluid particles inside the secondary vortices were noticed to oscillate. The frequency of the oscillations and their periodicity were also affected by the applied temperature gradients and the geometric shape of the injected bubbles. Recording the temperature fluctuations at a certain position near the injected air bubble can give a good indication about the periodicity of the oscillating liquid particles.

The main behaviour of temperature transient readings at a certain point during the current experiments can be explained as follows: for example, for a bubble aspect ratio $(\frac{r_B}{z_B})$ of 2.22, figure 4.3 and figure 4.4 show the temporal development of temperature readings at different locations near the bubble while pursuing our experiment.

First, low amplitude temperature variations of 0.03 °C before bubble injection ensure a stable vertical temperature gradient across the test cell as shown in figure 4.5 (a). These low amplitude temperature variations are caused by temperature sensor self-instabilities. Then, an instantaneous increase in temperature is observed, once the bubble is injected. Temperature oscillations, shown in figure 4.5 (b), with a low frequency of 0.1 Hz of temperature take about 7 minutes and then reach a plateau of 26.25 °C with very low amplitude temperature fluctuations which can be neglected as shown in figure 4.5 (c).

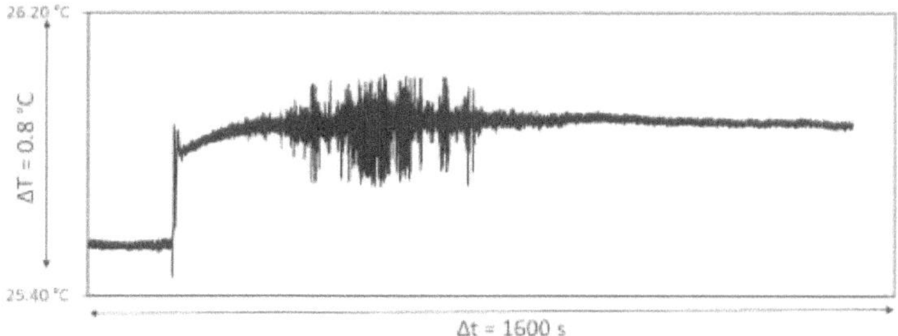

*Figure 4.3: Temperature fluctuation plot at another point 1.5 mm below point B in the liquid, $r_B$ = 6 mm, $z_B$ = 2.7 mm, $\frac{r_B}{z_B}$ = 2.22, $\Delta p$ = 0 bar, $\left|\frac{\partial T}{\partial z}\right|$ = 1.79 K/mm, and Mg = 43470*

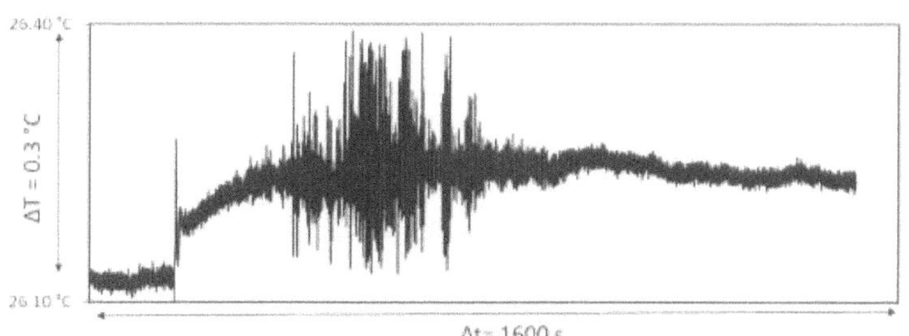

*Figure 4.4: Temperature fluctuation plot at point B in the liquid, $r_B$ = 6 mm, $z_B$ = 2.7 mm, $\frac{r_B}{z_B}$ = 2.22, $\Delta p$ = 0 bar, $\left|\frac{\partial T}{\partial z}\right|$ = 1.79 K/mm, and Mg = 43470*

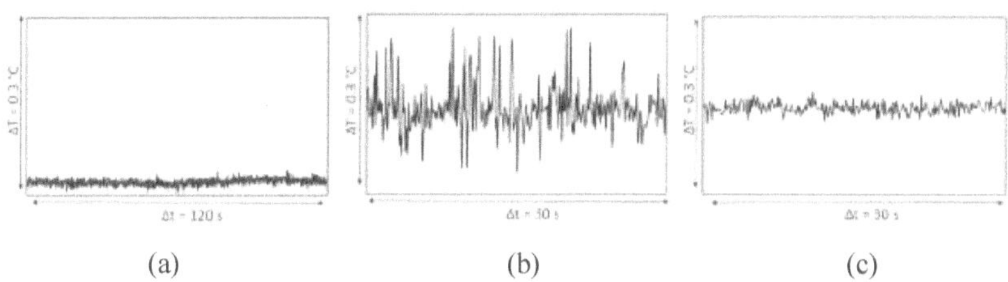

|        |        |        |
|:------:|:------:|:------:|
| (a)    | (b)    | (c)    |

*Figure 4.5: Temperature fluctuation plots at point B in the liquid, $r_B$ = 6 mm, $z_B$ = 2.7 mm, $\frac{r_B}{z_B}$ = 2.22, $\Delta p$ = 0 bar, $\left|\frac{\partial T}{\partial z}\right|$ = 1.79 K/mm, and Mg = 43470. (left: before bubble injection) (middle and right: after bubble injection)*

The frequency of the temperature fluctuations of 0.4 Hz, shown in figure 4.6 (upper), shows a satisfying agreement with the measured oscillating frequency of the fluid particles moving around the bubble. The mentioned oscillatory frequency of the fluid particles was determined by tracing the movement of a single particle around the injected air bubble. That was achieved with the help of 27 consecutive PIV camera shots of the flow field with a total time of 3.375 secs as shown in figure 4.6 (lower), where the tracer particles oscillate in the vertical direction (direction of the applied temperature gradient). It should be also noted that the PIV camera was adjusted with longer acquisition time to generate the tracer particle image shown in figure 4.6 (lower), where the tracer particles appeared larger than their normal size. Therefore, the combination of both PIV flow visualization and transient temperature measurements can give a satisfying qualitative and quantitative image of oscillations inside the flow field around the injected bubble.

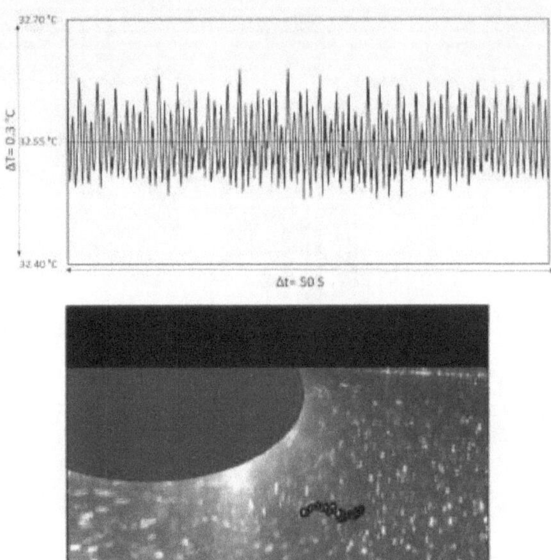

Figure 4.6: Temperature fluctuation plot at point B in the liquid (upper) and tracing of a single particle in the flow field around the bubble (lower) and, $r_B$ = 6 mm, $z_B$ = 2.7 mm, $\frac{r_B}{z_B}$ = 2.22, $\Delta p$ = 0 bar, $\left|\frac{\partial T}{\partial z}\right|$ = 1.79 K/mm, Mg = 43470, and f = 0.4 Hz

Generally, increasing the applied temperature gradients for the same bubble aspect ratio leads to a development of the thermocapillary flow from a steady laminar state to a periodic oscillatory state. To mention only one example, for a relatively large bubble aspect ratio of 2.24, while further increasing the applied temperature gradient along the test cell, the flow field around the bubble remained laminar till reaching Mg = 19500 at an applied temperature gradient of 0.8 K/mm. At this Mg-number, the first vortex was enlarged near the upper plate as shown in figure 4.7. The flow around the bubble was characterized by typical periodic temperature oscillations of a low frequency of 0.2 Hz. Furthermore, the small variations of amplitude, which are in a range of 0.03 °C are also neglected and considered as noise in the temperature readings.

The already mentioned case of periodically oscillating flow of Mg = 19500 matches the results of the study mentioned before by Raake et al. [22], where oscillations were supposed to arise starting from a temperature gradient of 0.5 K/mm, followed by a high degree of oscillations at a higher temperature gradient of about 1 K/mm. As shown later in figure 4.14, the current discussed case of Mg = 19500 and an injected air bubble of $\frac{r_B}{z_B}$ = 2.24 according to the experimental work of Chun et al. al [25] was expected to oscillate periodically. This matches our present results. The agreement on the value of critical Marangoni number, where the fluid flow around the bubble starts to oscillate, between our current study and past authors assures the validity of using the temperature plots to examine the periodicity of the flow oscillations. However, the current study does not focus on finding the transitional Mg-number ($Mg_{tran}$) for different bubble aspect ratios, as this point was well discussed in past studies.

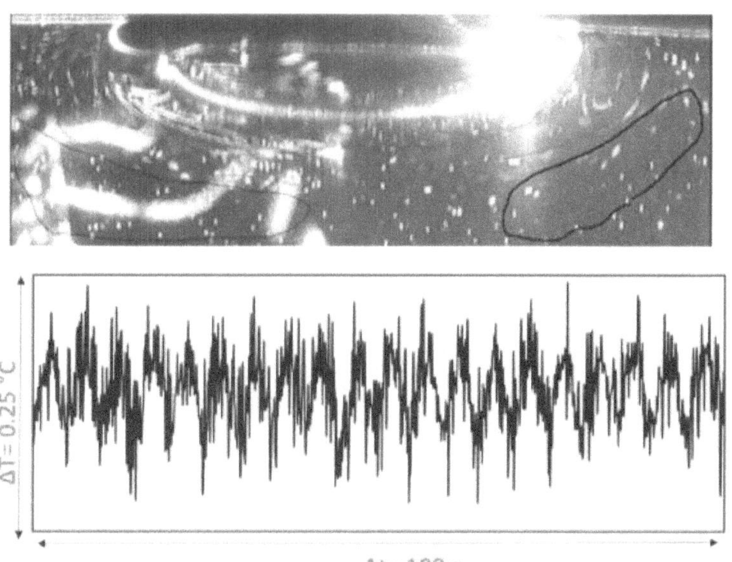



*Figure 4.7: Flow field around the bubble (upper) and temperature fluctuation plot at point B in the liquid (lower),*
$r_B$ = 6.05 mm, $z_B$ = 2.7 mm, $\frac{r_B}{z_B}$ = 2.24, Δp = 0 bar, $\left|\frac{\partial T}{\partial z}\right|$ = 0.8 K/mm, Mg = 19500, and f = 0.2 Hz

In the region of periodic oscillations, the periodicity of the fluid flow is not only examined through temperature fluctuations but also through the transient axial velocity measurements (in the direction of temperature gradient) at different positions around the bubble.

For a bubble aspect ratio of 1.3 and at a temperature gradient of 2.33 K/mm, one can conclude a periodic thermocapillary flow behaviour. This interesting control measurement showed, that the transient values of both the vertical velocity component at point A, located beneath the injected bubble (r = 1.6 mm, z = 4.25 mm), and the temperature measurements at point B, located beneath the upper plate but farther away from the bubble (r = 8.5 mm, z

= 4.0 mm), displayed the same approximate frequency of 0.8 Hz as shown in figure 4.8. The slight difference in the value of the frequency could be justified as both PIV frequency and temperature measurement frequency were generated from two separate experiments at the same boundary conditions as mentioned before. The temperature measurement uncertainty was ± 0.01 K, which is around 10 % of the displayed temperature fluctuations. The build-in temperature sensors were positioned about 8.5 mm radially away from the center line to enable experiments with larger bubbles. However, the PIV camera provided a smaller field of view beneath the upper plate for velocity measurements as mentioned before. That is why these velocity and temperature control measurements were conducted at different points.

It should also be noted that the flow velocity magnitudes around the bubble, shown in figure 4.8 (b), are not the maximum flow velocity as it is usually found on locations so near to the air bubble interface.

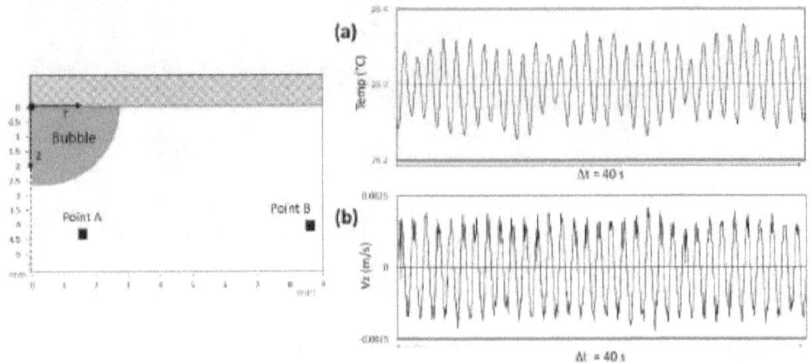



*Figure 4.8: (a): Typical temperature fluctuations at point B, (b): typical vertical velocity fluctuations at point A ($r_B$ = 3.2 mm, $z_B$ = 2.45 mm, $\frac{r_B}{z_B}$ = 1.31, $\Delta p$ = 0 bar, $\left|\frac{\partial T}{\partial z}\right|$ = 2.33 K/mm, Mg = 26800, and f = 0.8 Hz)*

Successive images of the fluid velocity vectors around the bubble, shown in figure 4.9, demonstrate that the periodic velocity and temperature fluctuations are resulting from the periodic movement of the primary vortex around the bubble. This vortex does not disintegrate but changes its size, initiates these oscillations of the thermocapillary boundary contour. These patterns are displayed to give an expression of the fundamental periodic oscillations flow velocity fields. It should also be noted that due to light reflections on the bubble surface, which were eliminated using an imaging post processing tool, the velocity vectors on the bubble periphery could not be generated. That justifies the lack of velocity vectors on the white zone around the bubble. The patterns of this recirculation zone repeated themselves in a periodic manner.

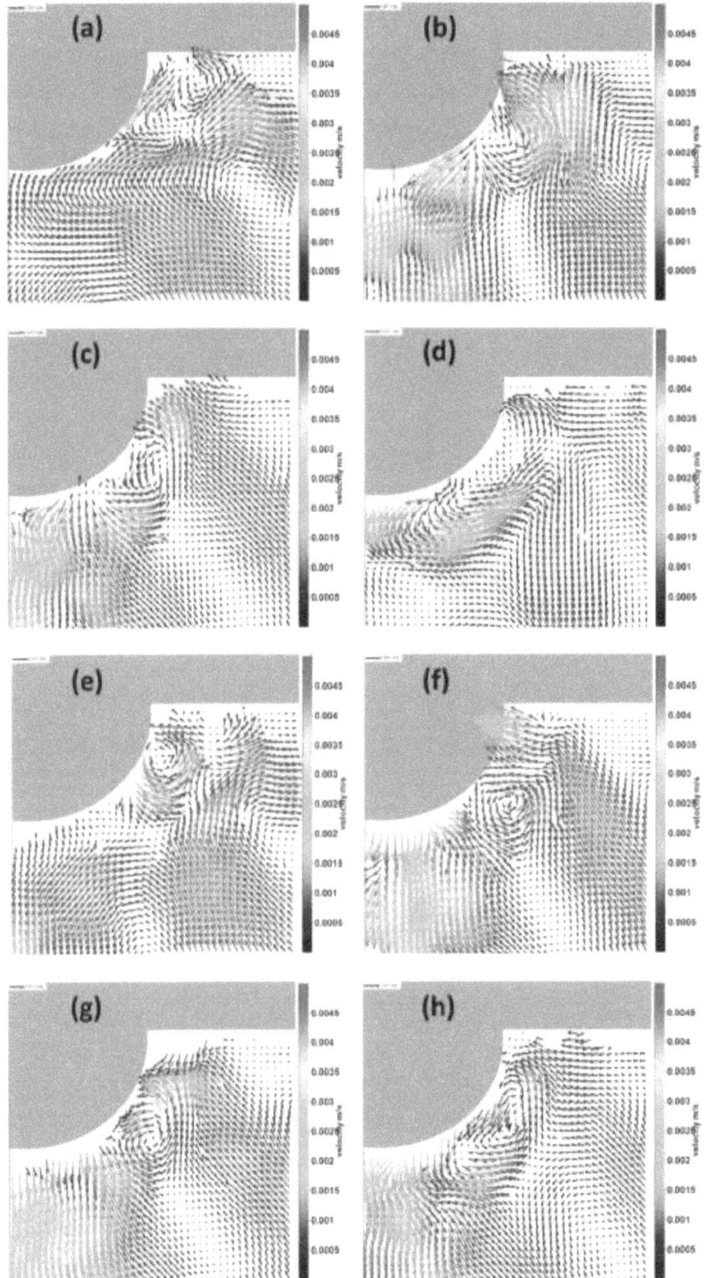

*Figure 4.9: PIV-measured velocity fields showing different positions of the recirculation zone around the air bubble, $r_B$ = 3.2 mm, $z_B$ = 2.45 mm, $\frac{r_B}{z_B}$ = 1.31, $\Delta p$ = 0 bar, $\left|\frac{\partial T}{\partial z}\right|$ = 2.33 K/mm, Mg = 26800, and f = 0.8 Hz*

In addition to that and for more accuracy, for a bubble aspect ratio of 1.44, transient velocity measurements at different points around the injected air bubble, shown in figure 4.10 and figure 4.11, show a good concurrency regarding the fluid flow oscillation frequency in the areas located beneath the thermocapillary main recirculation zone. Moreover, they assure the homogeneity of the fluid flow oscillations because of the thermocapillary bubble convection resulting from the applied vertical temperature gradient. These patterns of velocity vector field repeated themselves in a periodic manner with the same frequency of 0.82 Hz as observed in figure 4.12. Therefore, a homogeneous periodically oscillating fluid flow of frequency of around 0.82 Hz was observed when injecting an air bubble of an aspect ratio of 1.44 inside a AK 0.65 silicone oil fluid domain subjected to a vertical temperature gradient of 2.36 K/mm. Moreover, for bubbles of aspect ratios larger than 2.24, the periodic oscillatory state was expected to happen at temperature gradients less than 0.8 K/mm. In low Mg-number experiments, slightly above the critical Marangoni number ($Mg_c$), the effect of changing bubble diameter was also discussed. It was found that the duration of temperature fluctuations accompanied by bubble injection increases at larger bubbles; however, the span of temperature oscillations remained the same if the temperature gradient along the test cell is constant as shown in figure 4.13.

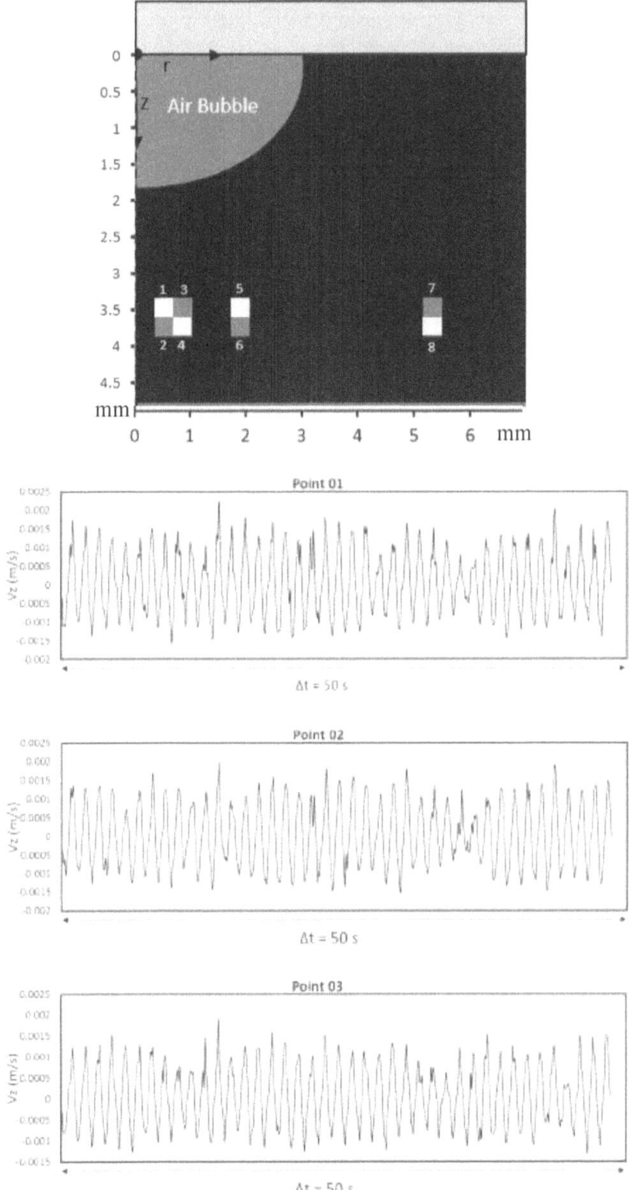


*Figure 4.10: Vertical velocity fluctuations at different points around the air bubble, $r_B = 3.65$ mm, $z_B = 2.53$ mm, $\frac{r_B}{z_B} = 1.44$, $\Delta p = 0$ bar, $\left|\frac{\partial T}{\partial z}\right| = 2.36$ K/mm, Mg $= 31850$, and $f = 0.82$ Hz*

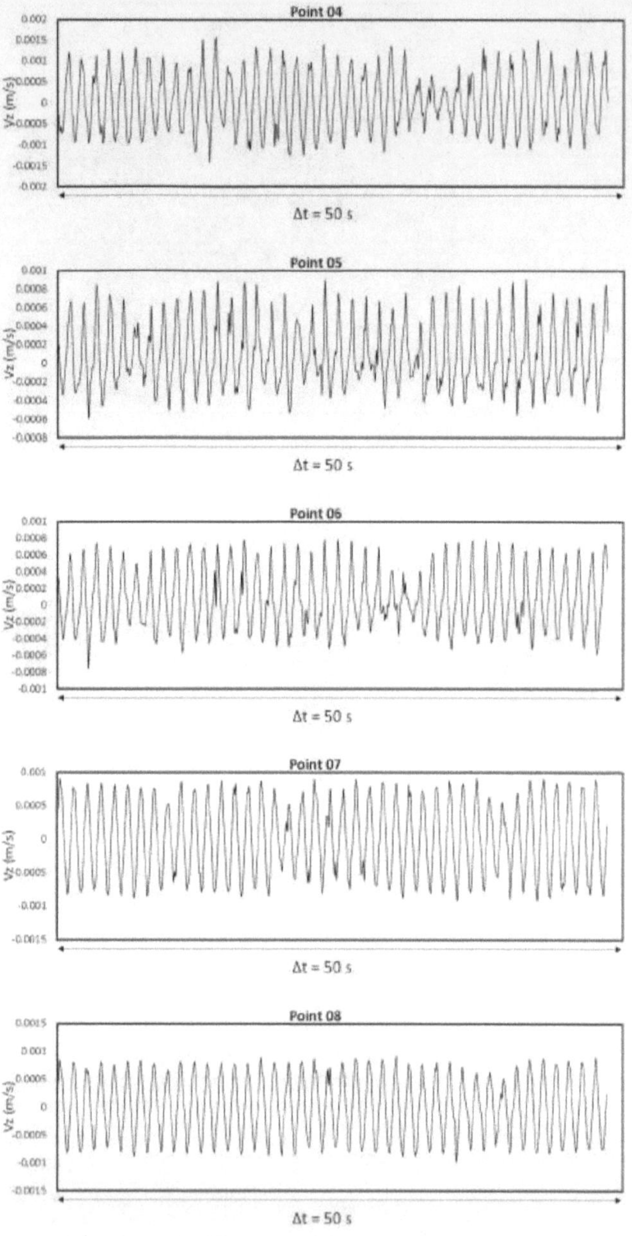

*Figure 4.11: Vertical velocity fluctuations at different points around the air bubble, $r_B = 3.65\ mm$, $z_B = 2.53\ mm$, $\frac{r_B}{z_B} = 1.44$, $\Delta p = 0\ bar$, $\left|\frac{\partial T}{\partial z}\right| = 2.36\ K/mm$, $Mg = 31850$, and $f = 0.82\ Hz$*

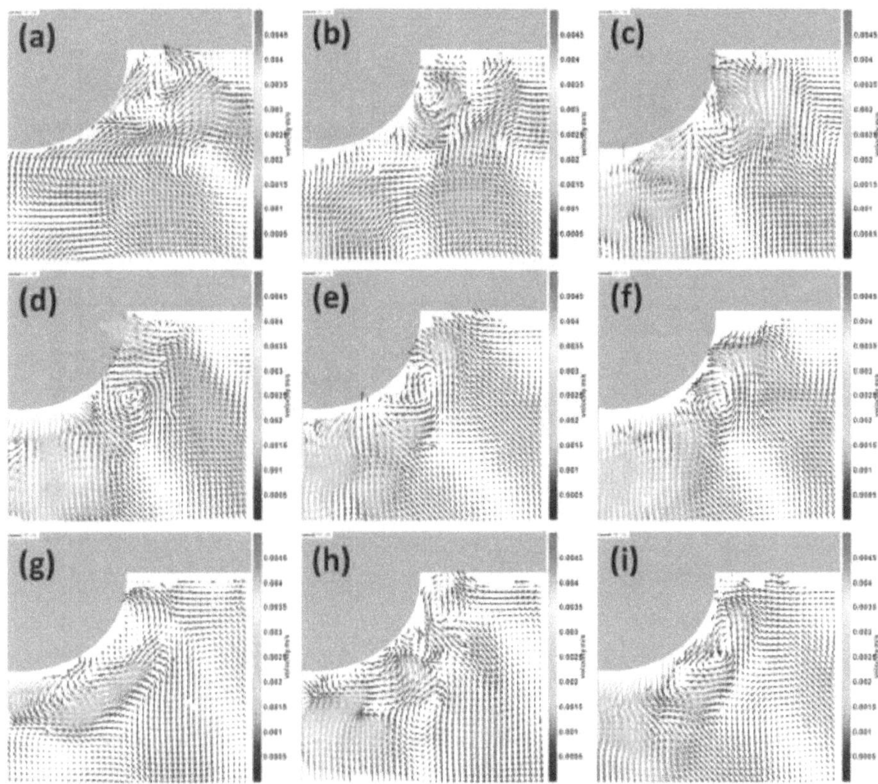

*Figure 4.12: PIV-measured periodic velocity fields, $r_B = 3.65$ mm, $z_B = 2.53$ mm, $\frac{r_B}{z_B} = 1.44$, $\Delta p = 0$ bar, $\left|\frac{\partial T}{\partial z}\right| = 2.36$ K/mm, Mg = 31850, and $f = 0.82$ Hz*

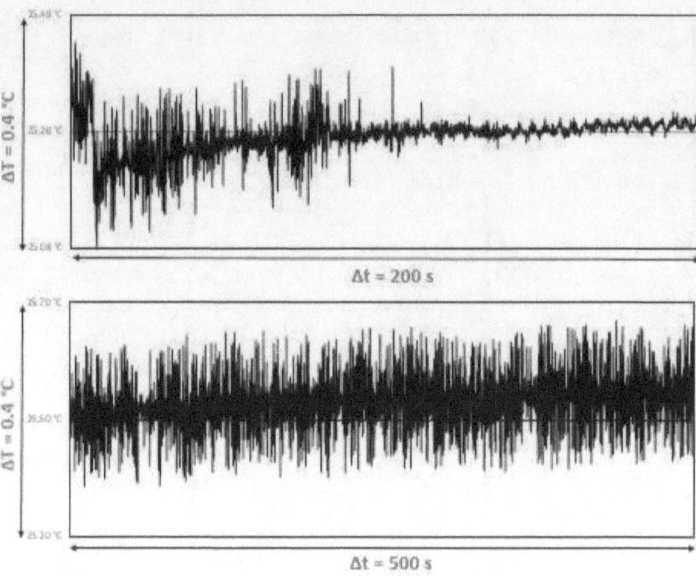



*Figure 4.13: Temperature fluctuation plots at point B in the liquid, $\left|\frac{\partial T}{\partial z}\right| = 0.7\ K/mm$, $\Delta p = 0\ bar$, (upper: $r_B = 7\ mm$, $z_B = 2.73\ mm$, $\frac{r_B}{z_B} = 2.56$, and Mg = 19130), (lower: $r_B = 8\ mm$, $z_B = 2.8\ mm$, $\frac{r_B}{z_B} = 2.86$, and Mg = 22430)*

Figure 4.14 shows that the current results using temperature measurements regarding the $Mg_{tran}$, where the periodic oscillations of the thermocapillary flow develops to a non-periodic state, are in good agreement with the past results [25]. The slight differences between the current and past results may be due to differences in the used temperature measuring techniques to define the effective temperature gradient near the bubble periphery. Also, the difference in the Prandtl number of the applied test liquid (current study: Pr = 5.2 to 7.2 and Chun et al. [25]: Pr = 5 to 6) could be a reasonable justification. It also shows that for bubbles of smaller aspect ratio, higher temperature gradients are needed to reach the zone of non-periodic oscillations.

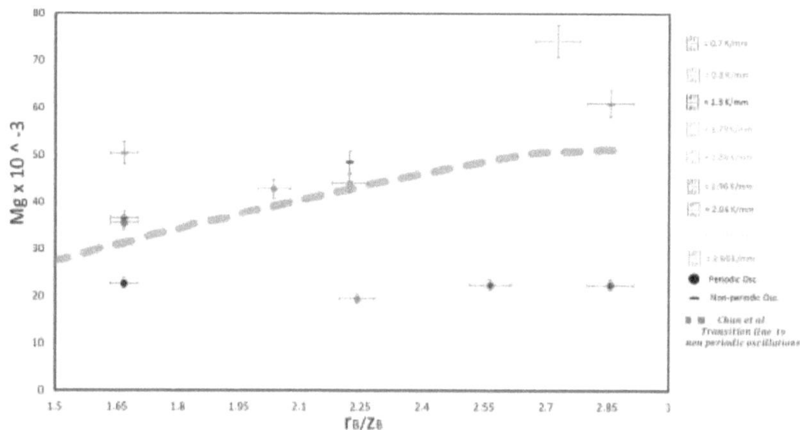

*Figure 4.14: Comparison between current temperature measurement results and Chuns et al. [25]*

As noticed in figure 4.15, at the same bubble aspect ratio $\frac{r_B}{z_B}$ of 2.24, the frequency of the periodic oscillations increases at higher temperature gradients. Moreover, for the same applied temperature gradient $\left|\frac{\partial T}{\partial z}\right| = 1.86$ K/mm, the frequency of periodic oscillations increases with bubbles of larger aspect ratios. However, for bubbles of aspect ratios more than 2.55, the frequency of periodic oscillations remains the same while increasing bubble aspect ratios. It means that the frequency of oscillations increases only when the applied increase in temperature gradient is sufficient to move the point to the region near the non-periodic oscillations as shown in figure 4.14.

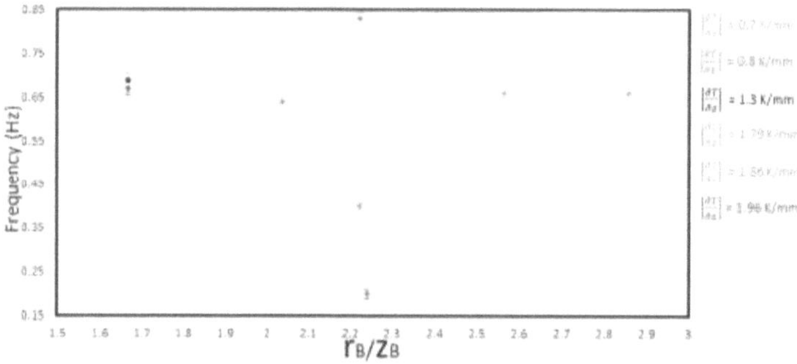

*Figure 4.15: Frequency of periodic oscillations using temperature measurements at different bubble sizes*

Further experiments are performed at higher Marangoni values. With higher temperature gradients, higher orders of temperature oscillations are noticed and a certain development in the flow field around the bubble is observed. The primary vortex splits into more eddies forming three vortices around the bubble, and the fluid tracer particles were found to be oscillating, which agrees with the findings of Raake et al. [22]. The temperature fluctuation plot, shown in figure 4.16, shows that the temperature fluctuations were periodic as expected but with a higher frequency of 0.64 Hz and a higher amplitude of 0.25 °C. This figure also shows that a sudden decrease followed by an increase in temperature was observed, which is quite strange. This sudden temperature jump repeated itself periodically every 35 seconds. That can be attributed to the periodic movement of the main vortex (shown in red arrows) in the vertical direction as shown in figure 4.16. It means that the movements of the main vortex disturb the stable temperature gradient around the bubble leading to those sudden increases in the observed temperature fluctuations. On the other side, unwanted mechanical vibrations of the injected bubble during the experiment may influence the recorded temperature fluctuations. Later on, using shadowgraphy (see chapter 4.2.2), that flow behaviour is attributed to the oscillations of the thermocapillary vortex boundary contour in both vertical and horizontal directions.

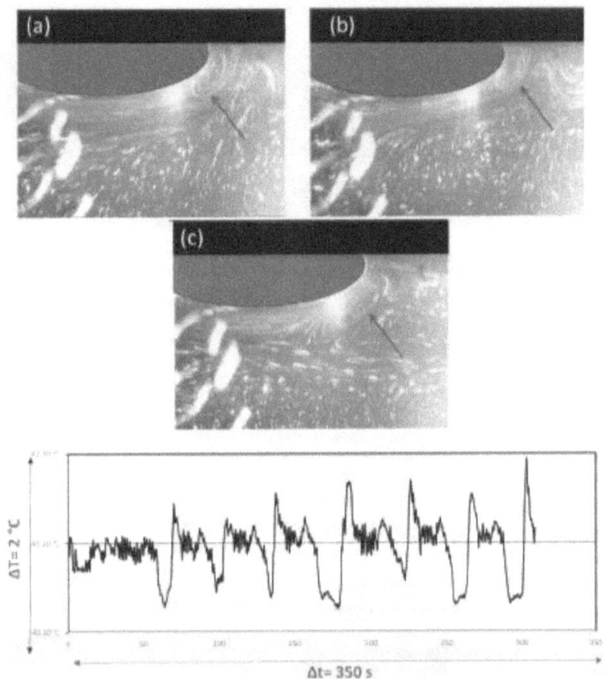



*Figure 4.16: Fluid flow field around the bubble, where red arrows show the different sizes and positions of the main vortex around the bubble (upper) and temperature fluctuations at point B in the liquid (lower), $r_B = 5.5$ mm, $z_B = 2.7$ mm, $\frac{r_B}{z_B} = 2.037$, $\Delta p = 0$ bar, $\left|\frac{\partial T}{\partial z}\right| = 1.86$ K/mm, and Mg = 42700*

### 4.1.2 Horizontal Light Sheet Results

As described before in chapter 3.7, consecutive imaging of the mentioned ring-shaped concentration of tracer particles in the horizontal plane provide a good description about the periodicity of the oscillatory thermocapillary-buoyancy fluid flow around the bubble as shown in figure 4.17.

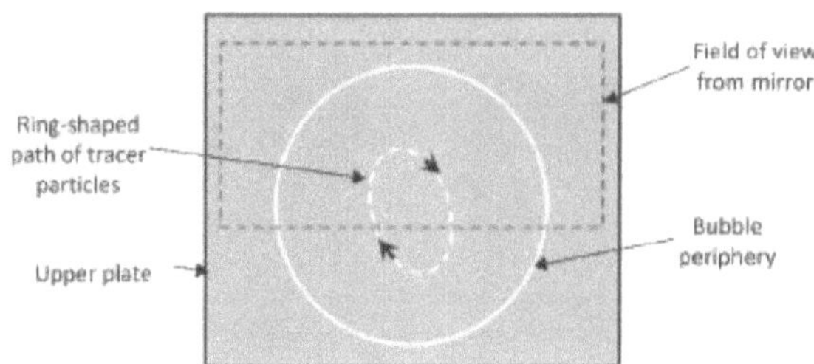


*Figure 4.17: Typical periodic circular rotation of an accumulation of tracer particles in the azimuthal plane*

Figure 4.18 shows a typical periodic circular rotation of the ring for a bubble aspect ratio of 1.67. That means that the flow is oscillating in a periodic manner. Further increase of the applied temperature gradient along the injected bubble enhances the Marangoni convective flow and accordingly influences the way, the tracer particles concentration ring moves. This enhancement appears as a movement of the center of rotation of the ring while rotating as shown in figure 4.19. This confirms the modification of the periodic oscillations to a higher mode.

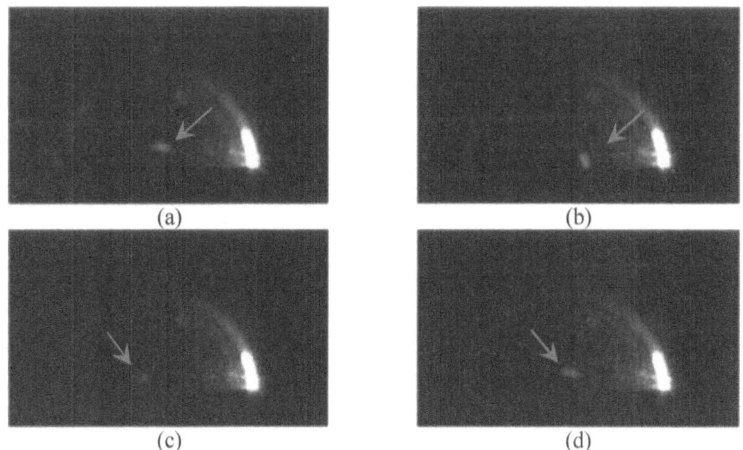


*Figure 4.18: Successive images of periodic oscillating flow, where red arrows show the transient positions of the ring-shaped tracer particles accumulation, $r_B = 4.5\ mm$, $z_B = 2.7\ mm$, $\frac{r_B}{z_B} = 1.67$, $\Delta p = 0\ bar$, $\left|\frac{\partial T}{\partial z}\right| = 1.3\ K/mm$, Mg = 22700 and $\Delta t = 0.25\ s$*

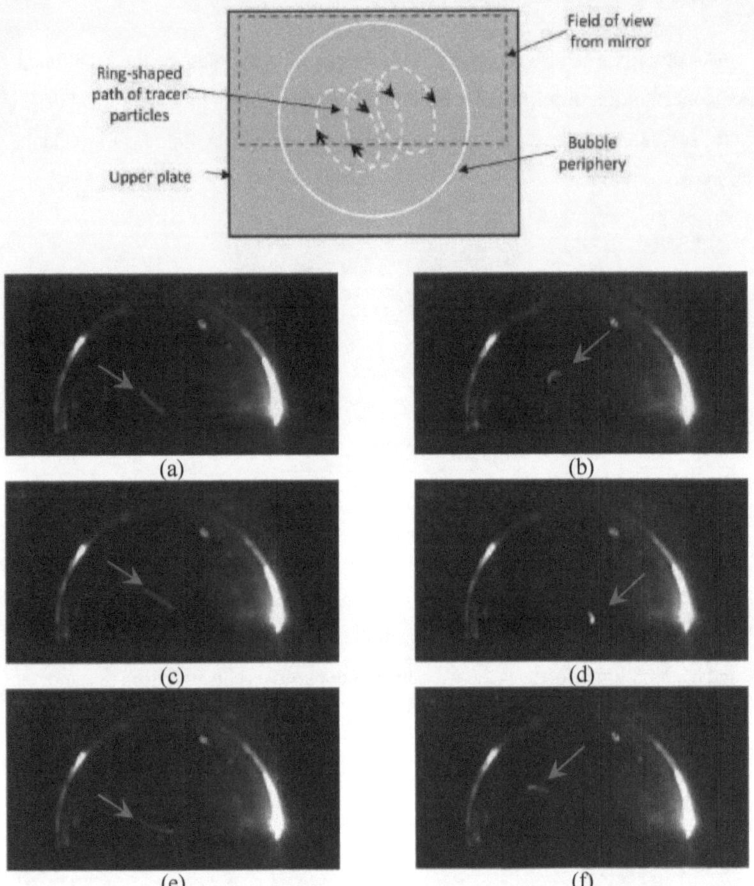



*Figure 4.19: Successive images of modified periodic oscillating flow, where red arrows show the transient positions of the ring-shaped tracer particles accumulation, $r_B = 4.5\ mm$, $z_B = 2.7\ mm$, $\frac{r_B}{z_B} = 1.67$, $\Delta p = 0\ bar$, $\left|\frac{\partial T}{\partial z}\right| = 2\ K/mm$, $Mg = 35420$ and $\Delta t = 0.5\ s$*

Comparing our temperature measurement results with our horizontal light sheet visualization technique results at the same bubble aspect ratio of 1.67 shows that temperature fluctuations remain periodic up to $\left|\frac{\partial T}{\partial z}\right| = 2$ K/mm. That can be also confirmed through the horizontal light sheet visualization technique (see figure 4.18 and figure 4.19). Then, at higher temperature gradients, the temperature fluctuations develop to reach a non-periodic state (see figure 4.20: b and c). Moreover, our value of the transitional Marangoni number, where the liquid flow reveals a non-periodic state, at an bubble aspect ratio of 1.67, agrees with the value of Chun et al. al [25]. That also validates and confirms the ability of the used measuring techniques to demonstrate the periodicity of the fluid flow around the injected bubble at different boundary conditions. A further increase of the applied temperature gradients leads to chaotic shapes of the path of tracer particles concentration in the horizontal plane which continue beyond the periodic oscillatory range of Mg-numbers. Tracing the movement of the tracer particles concentration below the bubble results in an incomprehensible image of the flow trajectories. Therefore, this methodology unfortunately did not help to analyze these chaotic movements of the tracer particles concentration and to describe properly the non-periodic state of the flow around the injected bubble at higher Mg-numbers.

(a)

(b)

(c)

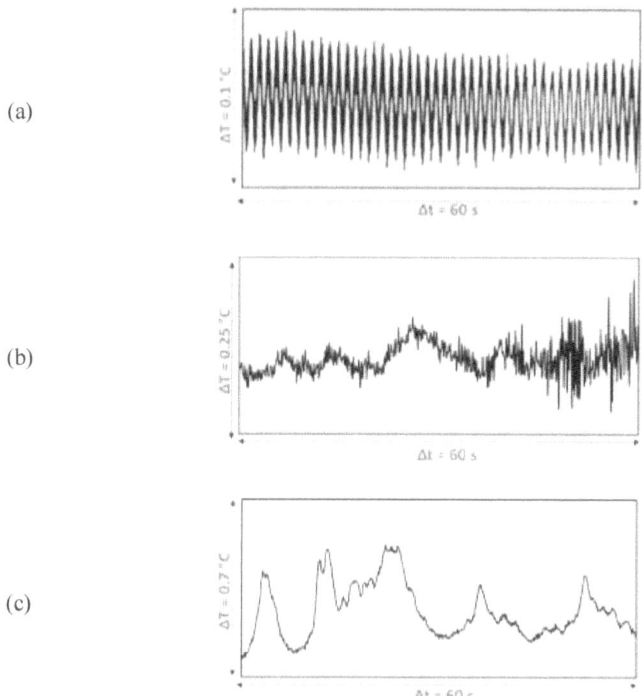


*Figure 4.20: Temperature fluctuation plots at point B in the liquid, $r_B$ = 4.5 mm, $z_B$ = 2.7 mm, $\frac{r_B}{z_B}$ = 1.67, (a): Mg = 35450, $\left|\frac{\partial T}{\partial z}\right|$ = 1.96 K/mm, p = 0 bar. (b): Mg = 36400, $\left|\frac{\partial T}{\partial z}\right|$ = 2.04 K/mm, Δp = 0 bar. (c): Mg = 50350, $\left|\frac{\partial T}{\partial z}\right|$ = 2.66 K/mm, ΔP = 1.5 bar*

### 4.1.3 Shadowgraphy Results

The main goal of this study is to focus on the non-periodic Marangoni convective flow behaviour which has not been studied in detail elsewhere so far. Therefore, some images of the periodic movement of the thermocapillary vortex boundary contour are presented to give an impression how it looks at relatively moderate Marangoni numbers. For visualizing the thermocapillary vortex boundary contour around the gas bubble we used a similar setup to the one used before by Xi et al. [49] to visualize the thermal plumes that initiate the turbulence within a Bénard-Rayleigh convection.

At moderate Marangoni numbers between 16000 and 27000 at different bubble sizes, the thermocapillary vortex boundary contour oscillates up and down in the vertical direction as shown in figure 4.21. This mode of oscillation is termed periodic in the vertical direction, where the thermocapillary vortex boundary contour mainly oscillates vertically (in the same direction of the applied temperature gradient) rather than the oscillations in the transverse direction (parallel to the upper heated wall). Accordingly, when using low-height test cells, to easily establish higher temperature gradients, the manner of the oscillation of the thermocapillary vortex boundary contour may be influenced by the lower wall. The lowest peak position of this boundary contour is noticed to be almost 80% of the bubble height. That means, by moderate bubble diameters of 8 mm and height of 2.6 mm, the thermocapillary vortex boundary contour oscillates with a vertical displacement of approximately 2.25 mm. Therefore, an enough minimum free space under this boundary contour of 10 mm is assumed to avoid any noticeable interactions between the solid boundary (bottom plate) and the oscillating thermocapillary flow. In the current study, the free height between the upper and bottom copper plate is kept at minimum value of 15 mm.

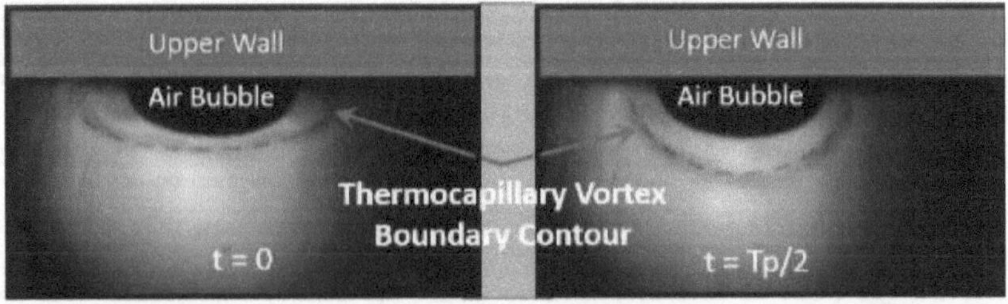



Figure 4.21: Periodic mode of oscillations of the thermocapillary vortex boundary contour in vertical direction, $r_B = 2.75\ mm$, $z_B = 2.3\ mm$, $\frac{r_B}{z_B} = 1.2$, $\Delta p = 0\ bar$, $\left|\frac{\partial T}{\partial z}\right| = 2.14\ K/mm$, $Mg = 19470$, $f = 0.8\ Hz$, and $Tp = 1.25\ s$, Tp denotes the periodic time of oscillations

The mentioned stretching of this boundary contour in the vertical direction is also accompanied by expansions and compressions from the sides as shown in figure 4.21. That corresponds to periodic oscillations of the thermocapillary flow around the injected bubble. At slightly higher Mg-numbers, the thermocapillary vortex boundary contour begins to oscillate in a modified manner, where a central downward peak expansion is noticed in the midline of the bubble as shown in figure 4.22 (c).

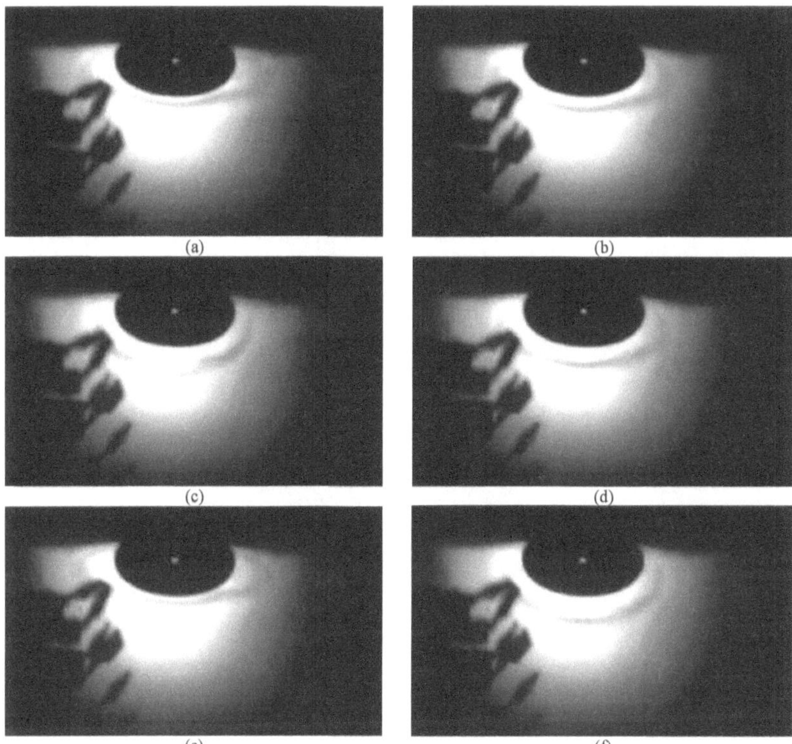

Figure 4.22: Periodic movement (modified oscillations) of the thermocapillary vortex boundary contour, $r_B = 3.5$ mm, $z_B = 2.5$ mm, $\frac{r_B}{z_B} = 1.4$, $\Delta p = 0$ bar, $\left|\frac{\partial T}{\partial z}\right| = 2$ K/mm, Mg = 25700, f = 0.75 Hz and $\Delta t = 1.67$ s

Moreover, the compatibility between the thermocapillary vortex boundary contour behaviour through shadowgraphy and the fluid particles movements through PIV can give a good image of the thermocapillary flow behaviour on both macro and micro scales. Figure 4.23 (upper) demonstrates the transient positions of the thermocapillary vortex boundary contour around a bubble of $\frac{r_B}{z_B} = 1.31$. It also shows that the oscillations in the vertical direction are dominant. Through consecutive shadowgraph images, these oscillations were observed to be periodic. Also, the transient PIV- measured velocity fields, displayed in figure 4.23

(middle), resulting from recording of tracer particles at the same boundary conditions, show that the movements of the primary vortex and the variety of its size initiated the mentioned oscillations of the thermocapillary boundary contour.

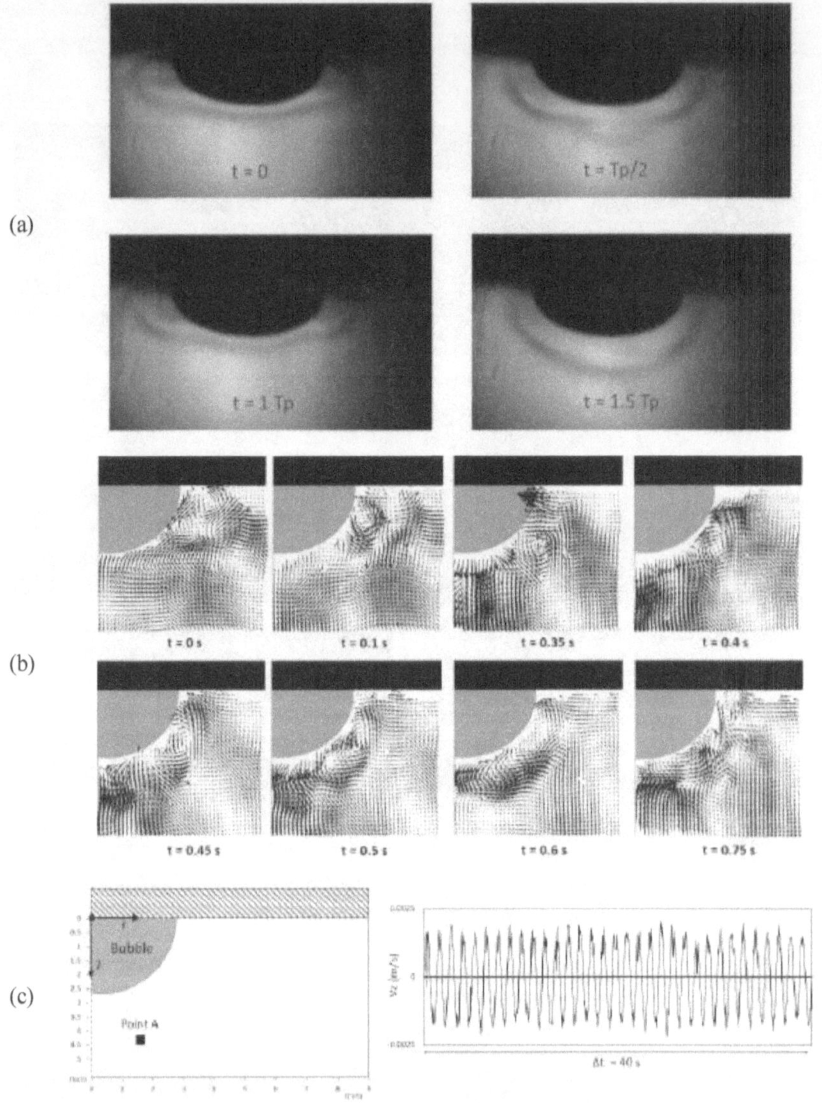

Figure 4.23: Comparison between upper: shadowgraph, middle: PIV-measured velocity fields, lower: velocity measurements ($v_z$) at point A, $r_B = 3.2$ mm, $z_B = 2.45$ mm, $\frac{r_B}{z_B} = 1.31$, $\Delta p = 0$ bar, $\left|\frac{\partial T}{\partial z}\right| = 2.33$ K/mm, Mg = 26800, Tp = 1.2 s and f = 0.8 Hz

Additionally, these patterns of the recirculation zone repeated themselves in a periodic manner, which confirms the periodicity of the flow oscillations. Through comparing both shadowgraphy and PIV results, the same frequency of 0.8 Hz was confirmed. Also, interestingly, figure 4.23 shows that the frequency of the vertical velocity component at point A, located beneath the injected bubble (r = 1.6 mm, z = 4.25 mm), concurs well with the frequency concluded from shadowgraphy. This good agreement between qualitative results through shadowgraphy and quantitative results through velocity and temperature measurements demonstrates how far the shadowgraph images can detect the thermocapillary flow behaviour in the current flow configuration.

Generally, the current shadowgraphy results within the zone of periodic oscillations and our values of Mg$_{tran}$ for different bubble aspect ratios show satisfying agreement with previous results of Chun et al. [25] as shown in figure 4.24. Slight differences between our values of Mg$_{tran}$ and theirs may result from differences in the temperature measuring techniques or the criteria used by Chun et al. [25] to calculate the applied temperature gradient as mentioned before. Figure 4.24 also shows that for bubbles of smaller aspect ratio, a higher temperature gradient is needed to reach the zone of non-periodic oscillations.

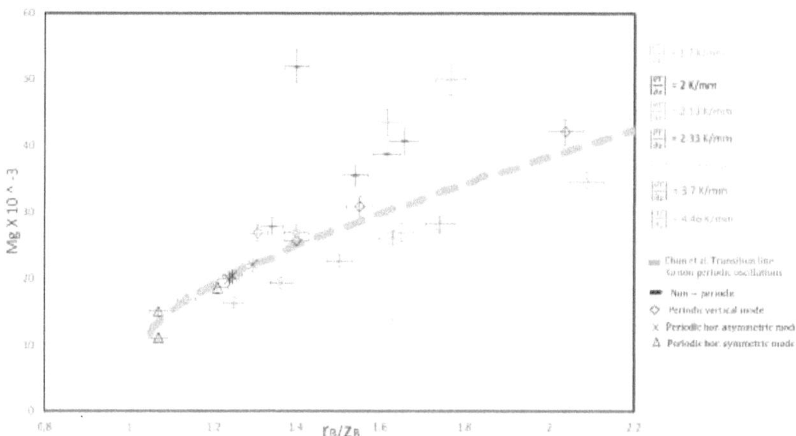


*Figure 4.24: Comparison between current shadowgraphy results and results of Chun et al. [25]*

Other modes of periodic oscillatory flow behaviour around the gas bubble were also noticed. These modes were distinguished from the previous mentioned mode of periodic oscillations in the vertical direction as the boundary contour was observed to oscillate mainly in the horizontal direction as shown in figure 4.25, where both symmetric and asymmetric oscillations in the transverse direction are dominant.

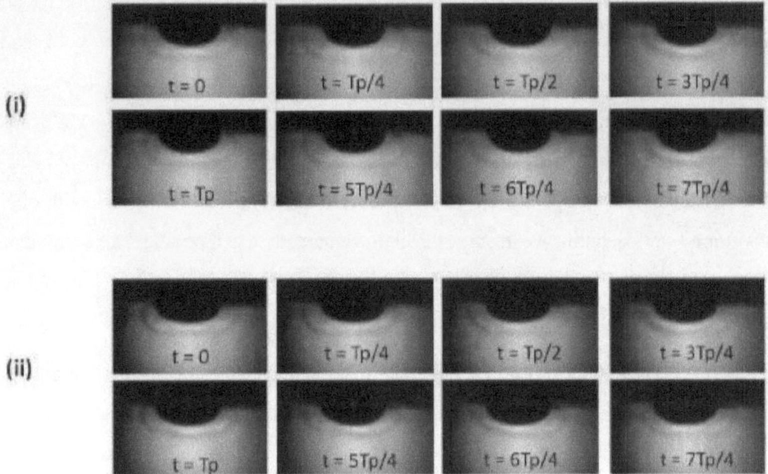



*Figure 4.25: Different periodic thermocapillary vortex boundary contour oscillation modes in the horizontal direction through shadowgraph images (a: symmetric oscillations; b: asymmetric oscillations), (i): $r_B = 2.35$ mm, $z_B = 2.2$ mm, $\frac{r_B}{z_B} = 1.07$, $\Delta p = 0$ bar, $\left|\frac{\partial T}{\partial z}\right| = 2$ K/mm, Mg = 15049, $\Delta t = 7$ s and $f = 0.2$Hz and (ii): $r_B = 3$ mm, $z_B = 2.4$ mm, $\frac{r_B}{z_B} = 1.25$, $\Delta p = 0$ bar, $\left|\frac{\partial T}{\partial z}\right| = 1.6$ K/mm, Mg = 26800, $\Delta t = 6.25$ s and $f = 0.28$ Hz*

These modes of symmetric and asymmetric horizontal periodic oscillations were observed at low temperature gradients from 1 K/mm to 2 K/mm. The oscillation frequencies of these modes are of values between 0.2 to 0.4 Hz. Additionally, the vertical mode of periodic oscillations was usually observed at higher temperature gradients just below the transitional border to the zone of non-periodic oscillations. Compared to the frequencies of horizontal oscillations, the frequencies of the vertical mode oscillations are in a range of 0.6 to 0.85 Hz. Moreover, these relatively higher frequency vertical periodic oscillations were dominant at temperature gradients above 2 K/mm regardless of the bubble sizes (see figure 4.26). Accordingly, it is easily concluded that at each bubble aspect ratio, the oscillations start with the low frequency mode of periodic oscillations in the horizontal direction. Then, with increasing the applied temperature gradient, the periodic mode of oscillations in the vertical direction appears till reaching the Mg$_{tran}$ and then, the thermocapillary flow develops to a non-periodic state.

In other words, using shadowgraphy, a concrete estimation for the values of the Mg$_{tran}$ at three different bubble sizes was concluded as shown in figure 4.27. It also shows that the transitional temperature gradient, where non-periodic oscillations start, is inversely proportional to the bubble aspect ratio.

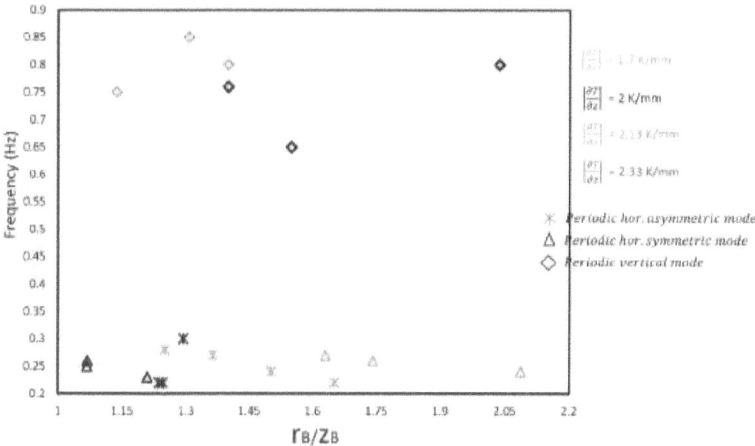

Figure 4.26: Frequency of periodic oscillations using shadowgraphy at different bubble sizes and temperature gradients

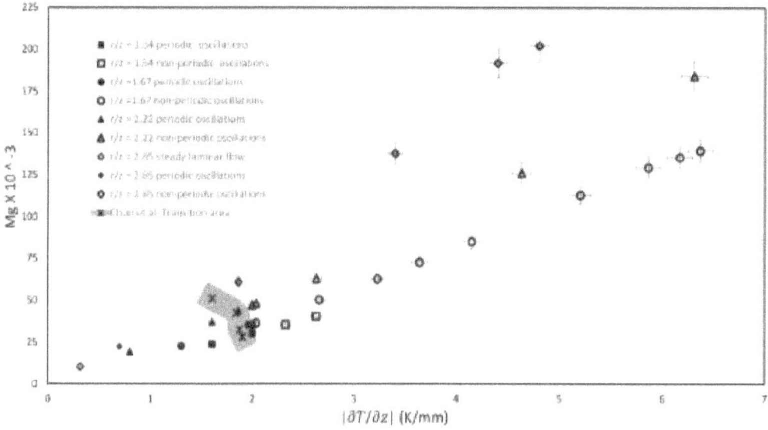

Figure 4.27: Transition of thermocapillary bubble flow towards non-periodic oscillations compared to the results of Chun et al. [25] for different bubble aspect ratios ($\frac{r_B}{z_B}$ = 1.54, 1.67, 2.22, and 2.85)

## 4.2  Non-Periodic Thermocapillary Flow

### 4.2.1 PIV and Temperature Measurement Results

As mentioned before (see chapter 3.3.4), PIV tracer particle images of the experiments conducted inside the pressure chamber were not adequate to the rapid flow velocities within the non-periodic flow regime compared to shadowgraphy. The rapid chaotic movements of tracer particles at conditions beyond the critical Mg-number needed a higher capturing frequency and accordingly a shorter acquisition time ($t_{acq}$) for every image. In general, the thermocapillary bubble flow within the non-periodic oscillation region is characterized by rapid movements of the primary recirculation zones in the transverse direction parallel to the upper horizontal wall. These movements are also accompanied by chaotic oscillations of the tracer particles beneath the bubble and the recirculation zone. Moreover, to focus on the elementary oscillations of the fluid particles in a micro scale, tracer particle visualization technique was also used but with longer acquisition time, where fluid flow trajectories could be captured as shown in figure 4.28. That confirms the non-periodic flow behaviour but cannot be evaluated quantitatively.

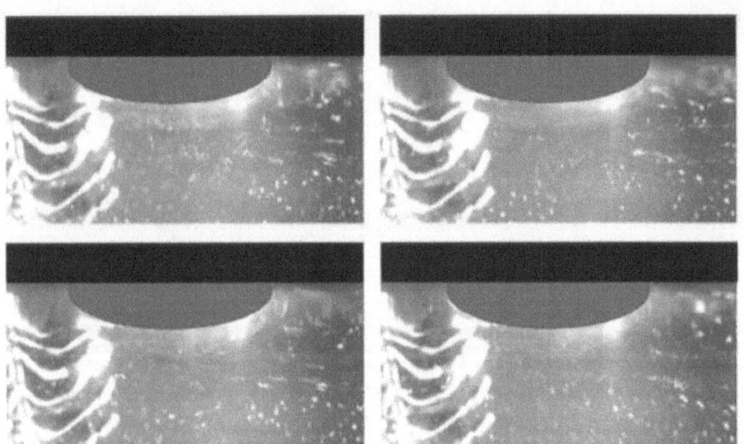

*Figure 4.28: Chaotic trajectories within the flow field around the right side of the injected bubble, $r_B = 6$ mm, $z_B = 2.7$ mm, $\frac{r_B}{z_B} = 2.22$, $\Delta p = 0$ bar, $\left|\frac{\partial T}{\partial z}\right| = 4.96$ K/mm, $t_{acq} = 2$ sec, and Mg = 135400*

As shown before in figure 4.27, our results of the transition of the thermocapillary bubble flow into a non-periodic state for different bubble sizes were compared to the results of Chun et al. [25]. The temperature fluctuation plots of the current experiments are shown in figure 4.29 and figure 4.30. These current temperature results, for every bubble aspect ratio, show that the values of temperature gradients required to generate non-periodic temperature fluctuations depend on the size of the injected gas bubble (bubble aspect ratio). These figures show obviously how temperature fluctuations develop by increasing the applied temperature gradient. On the same side, through these temperature fluctuation plots, the non-periodic behaviour of oscillations can be easily distinguished from the periodic one. Moreover, at relatively very high Mg-numbers, the temperature fluctuations were observed to alternate in a wide range of temperatures as shown in figure 4.31 rather than at lower Mg-numbers. That reveals non-periodic or turbulent behaviour of the thermocapillary bubble flow but could not quantify the turbulent state.

(a)

(b)

(c)

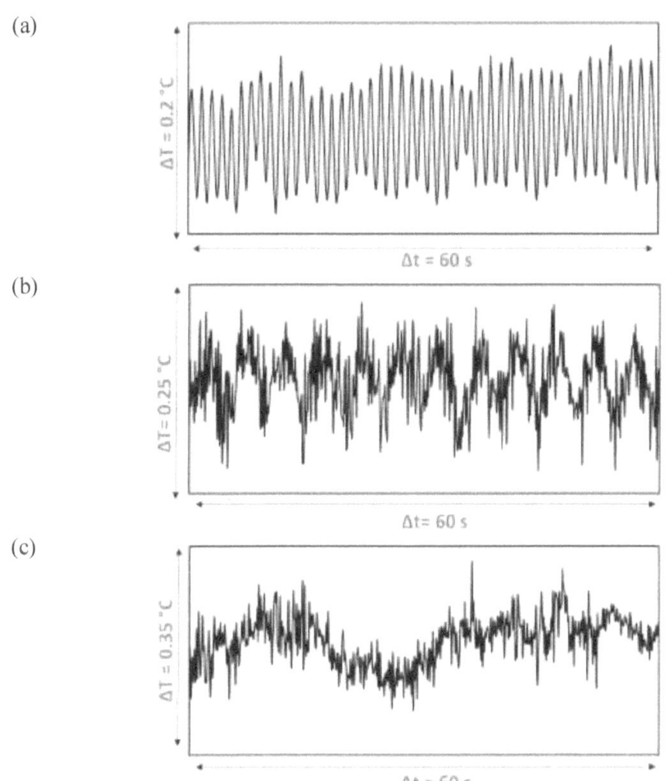

*Figure 4.29: Temperature fluctuation plots at point B in the liquid, $r_B = 6$ mm, $z_B = 2.7$ mm, $\frac{r_B}{z_B} = 2.22$, $\Delta p = 0$ bar.*
*(a): periodic oscillations at Mg = 19500, $\left|\frac{\partial T}{\partial z}\right| = 0.8$ K/mm, f = 0.2 Hz. (b): higher frequency periodic oscillations at*
*Mg = 44000, $\left|\frac{\partial T}{\partial z}\right| = 1.86$ K/mm, f = 0.83 Hz. (c): non-periodic oscillations at Mg = 48550, $\left|\frac{\partial T}{\partial z}\right| = 2.04$ K/mm*

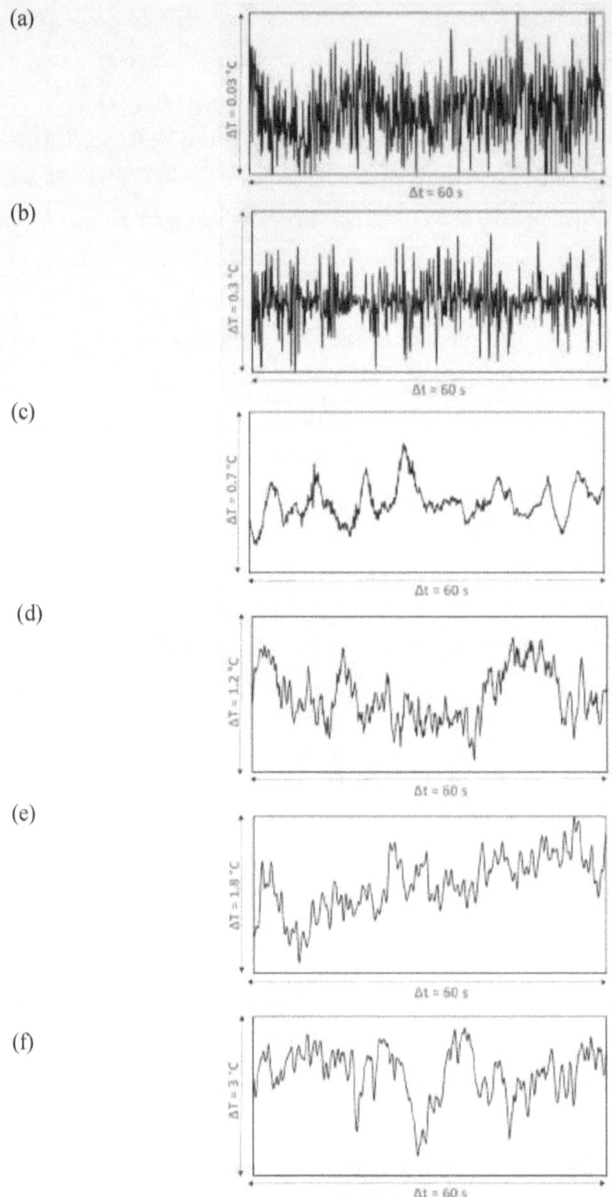

*Figure 4.30: Temperature fluctuation plots at point B in the liquid, $r_B = 8$ mm, $z_B = 2.8$ mm, $\frac{r_B}{z_B} = 2.85$, (a): laminar steady flow at $Mg = 10320$, $\left|\frac{\partial T}{\partial z}\right| = 0.32$ K/mm, $\Delta p = 0$ bar. (b): periodic oscillations at $Mg = 22430$, $\left|\frac{\partial T}{\partial z}\right| = 0.7$ K/mm, $\Delta p = 0$ bar, $f = 0.66$ Hz. (c): non-periodic oscillations at $Mg = 60900$, $\left|\frac{\partial T}{\partial z}\right| = 1.86$ K/mm, $\Delta p = 0$ bar. (d): non-periodic oscillations at $Mg = 137700$, $\left|\frac{\partial T}{\partial z}\right| = 3.4$ K/mm, $\Delta p = 1$ bar. (e): non-periodic oscillations at $Mg = 192900$, $\left|\frac{\partial T}{\partial z}\right| = 4.4$ K/mm, $\Delta p = 1$ bar. (f): $Mg = 202300$, $\left|\frac{\partial T}{\partial z}\right| = 4.8$ K/mm, $\Delta p = 1.3$ bar*

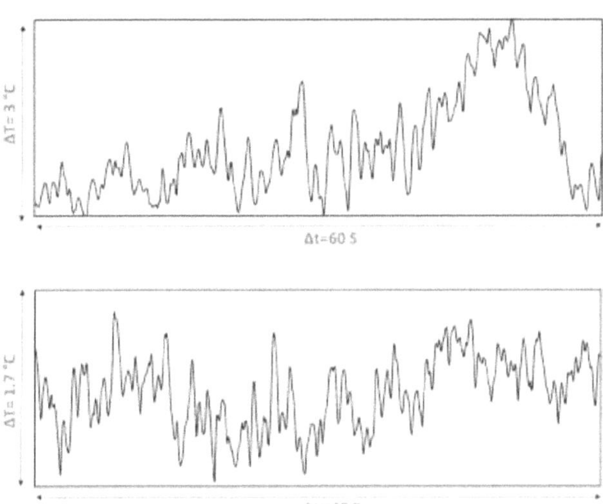


Figure 4.31: Turbulent temperature fluctuation plots at point B in the liquid, $r_B$ = 7.5 mm, $z_B$ = 2.75 mm, $\frac{r_B}{z_B}$ = 2.73, (upper: Mg = 159640, $\left|\frac{\partial T}{\partial z}\right|$ = 4.33 K/mm, $\Delta p$ = 0.7 bar) and (lower: Mg =189160, $\left|\frac{\partial T}{\partial z}\right|$ = 4.86 K/mm, $\Delta p$ = 1.3 bar)

In addition to temperature measurements, fluid flow velocities around the injected bubble gave us a good overview about the thermocapillary flow behaviour in the zone of non-periodic oscillations.

Using PIV and regardless of the expected higher values of the thermocapillary flow velocities, the fluid flow velocities around the bubble were measured at higher Mg-numbers through the evaluation of tracer particles images. At about Mg > 65000 and bubble aspect ratios between 1.0 and 2.0, the fluid flow velocities near the bubble periphery were measured. Due to relatively high velocities, higher frequency imaging with shorter capturing time was needed to adjust the PIV measurements. High frequency imaging requires a more powerful light source particularly when using the long-distance microscope. However, in our experiments the maximum imaging frequency was about 50 Hz, where the tracer particles near the bubble interface displayed short lines instead of the default dot-shaped field. Therefore, the velocities near the bubble periphery were calculated manually with the same criteria of the displaced distance between two successive images. The velocity measurement uncertainty was approximately ± 5 %.

The fluid flow is usually initiated through surface tension gradients and that can be clearly demonstrated through fluid flow velocity measurement, where the maximum fluid flow velocity magnitude ($|U|_{max}$) was detected close to the bubble interface approximately 0.2 mm away from the bubble surface. This thermocapillary flow driving velocity increases with the increase of the applied temperature gradients as shown in figure 4.32. Additionally, for a constant temperature gradient, higher velocities were found at larger bubbles.

The thermocapillary flow field is strongly influenced by the value of the thermocapillary flow driving velocity. First, as a result of the surface tension gradient on the bubble interface, the fluid particles on the bubble interface accelerate. After that, they decelerate until they stop their motion at the pole of the bubble to reverse their movement upwards amplified by buoyancy.

The magnitude of the maximum fluid flow velocity can reach a value of the order of 3 cm/s, when subjected to high temperature gradients. At these conditions, the fluid particles on the bubble interface decelerate faster to enable reversing their movements at the end of the bubble interface. The chaotic behaviour of the thermocapillary vortex boundary contour oscillations, shown in figure 4.33, is a result of this fast deceleration of the particles, which leads finally to chaotic fast movements and interactions between flow vortices.

As shown in figure 4.34, the turbulent oscillating behaviour of the thermocapillary flow around the injected bubble at high Mg-numbers results in oscillations in both the tracer particles and the thermocapillary vortex boundary contour (a`-c`). The different modes of non-periodic flow movements around the bubble are more comparable by tracing the movement of the thermocapillary vortex boundary contour through shadowgraph images rather than by tracing the elementary tracer particle movements through PIV images. Therefore, in the described flow configuration, shadowgraphy is the most adequate technique to capture comparable images of the non-periodic flow behaviour.

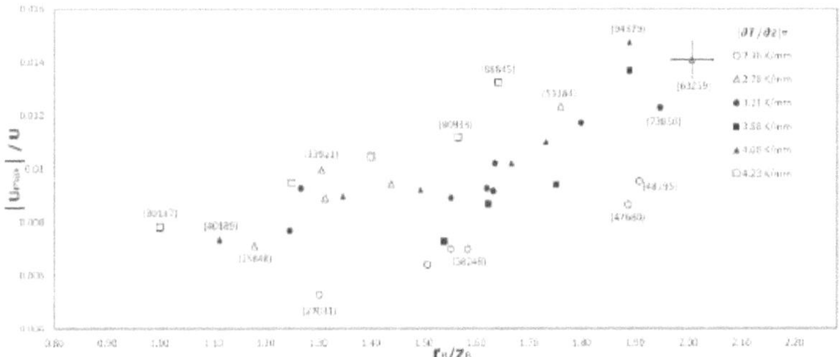

*Figure 4.32: Variation of the maximum flow velocity at the bubble surface with bubble shape at different temperature gradients. $|U|_{max}$ is scaled with the characteristic flow velocity U. Parameters in brackets are Mg-number values*

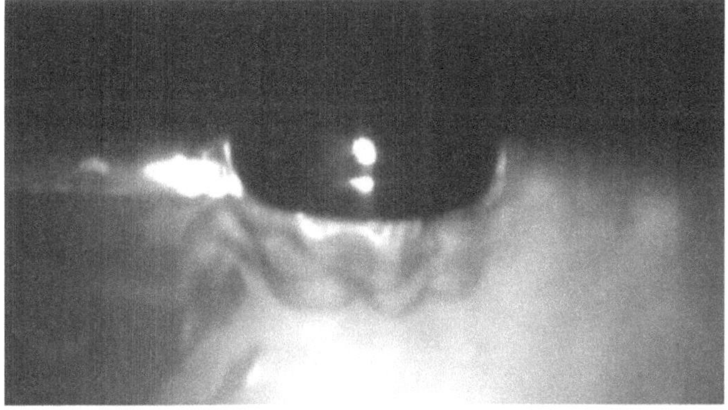

*Figure 4.33: Turbulent behaviour of the vortex boundary contour, $r_B = 4.41\ mm$, $z_B = 2.67\ mm$, $\frac{r_B}{z_B} = 1.65$, $\Delta p = 1.9\ bar$, $\left|\frac{\partial T}{\partial z}\right| = 4.96\ K/mm$ and $Mg = 101660$*

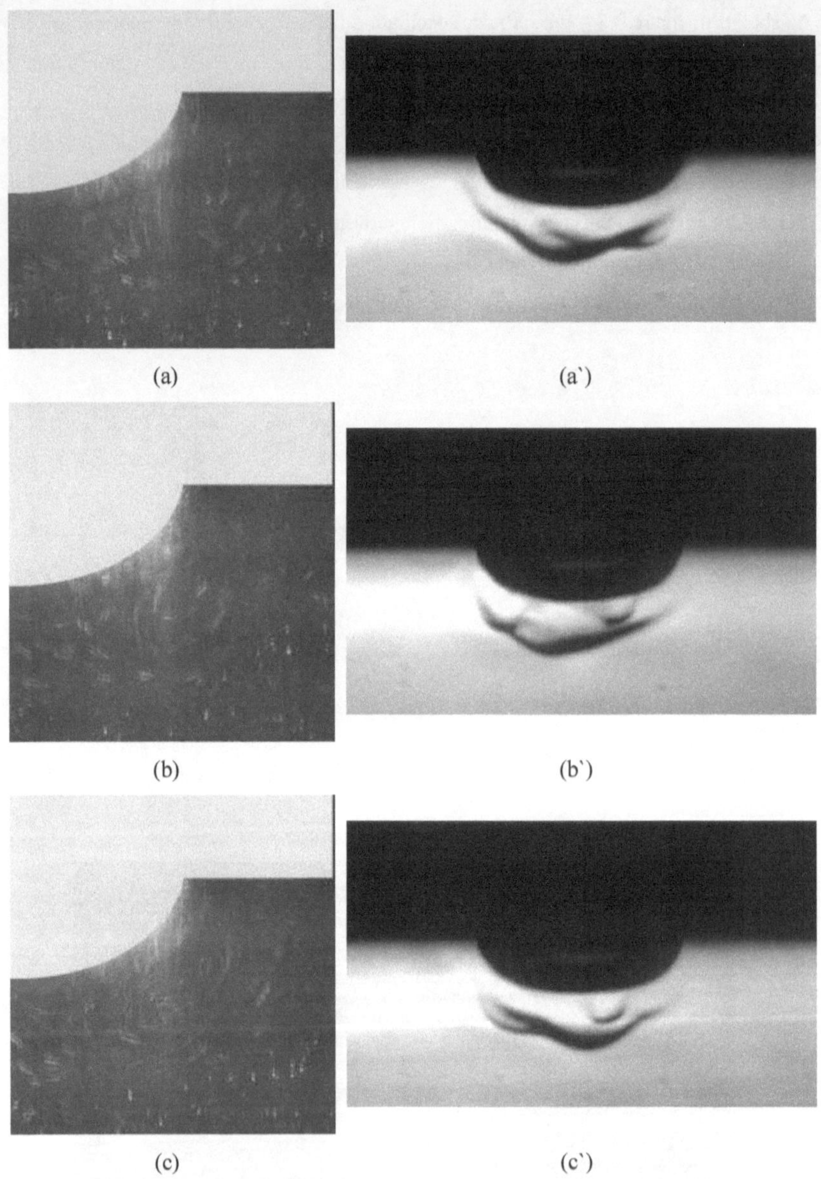

*Figure 4.34: PIV (a-c) and Shadowgraph images (a`-c`) for the same Mg-number (Mg = 70000)*

### 4.2.2 Shadowgraphy Results

Unlike PIV images, shadowgraph images are more applicable to evaluate the non-periodic flow behaviour around the bubble at high Mg-numbers by tracing the thermocapillary vortex boundary contour oscillations. At Mg-numbers higher than the value of $Mg_{tran}$ (see figure 4.35 and figure 4.36) and at each bubble aspect ratio, the boundary contour began to oscillate in a non-periodic manner, where extra downward peak expansions were observed in the transverse direction as shown in figure 4.37 and figure 4.38. These peak expansion amplitudes repeated themselves in a non-periodic manner.

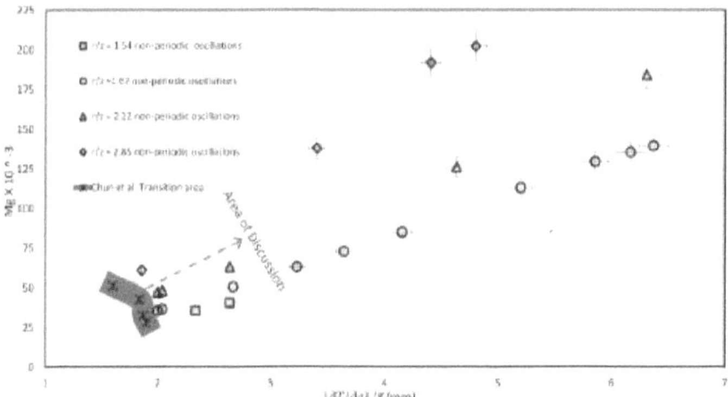

Figure 4.35: Non-Periodic thermocapillary flow oscillations area of interest

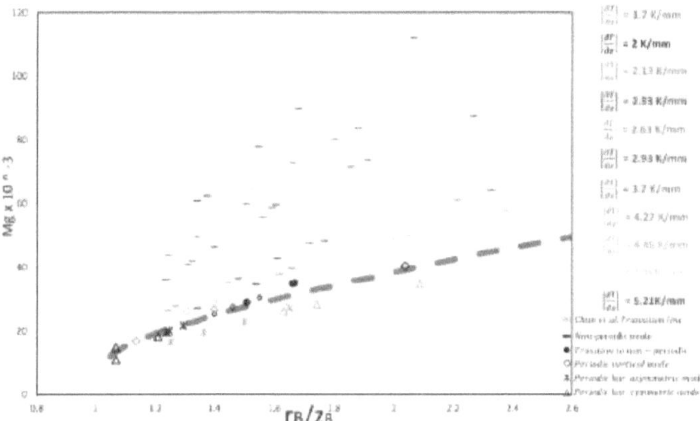

Figure 4.36: Variation of Mg with bubble aspect ratio for different experiments within both periodic and non-periodic oscillation zones

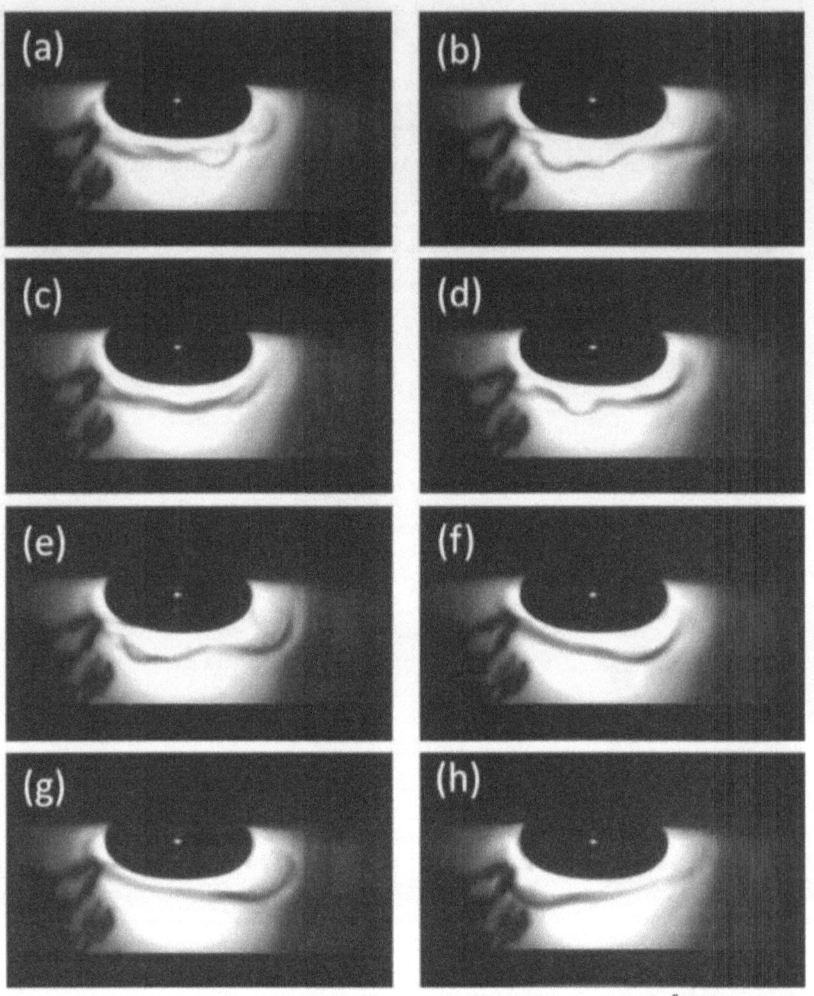

Figure 4.37: *Non-Periodic movements of vortex boundary contour*, $r_B = 4.23$ mm, $z_B = 2.66$ mm, $\frac{r_B}{z_B} = 1.61$, $\Delta p = 0$ bar, $\left|\frac{\partial T}{\partial z}\right| = 2.33$ K/mm, $Mg = 38911$ and $\Delta t = 5$ s

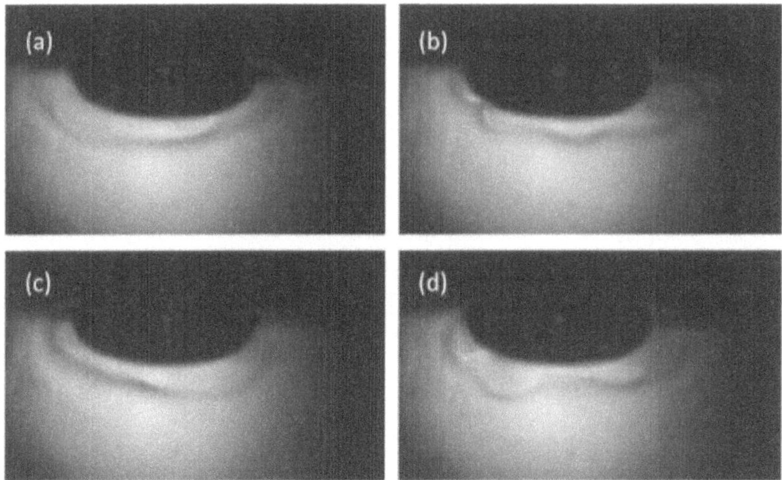

*Figure 4.38: Non-Periodic movements of vortex boundary layer, $r_B$ = 4.44 mm, $z_B$ = 2.68 mm, $\frac{r_B}{z_B}$ = 1.66, $\Delta p$ = 0 bar, $\left|\frac{\partial T}{\partial z}\right|$ = 2.33 K/mm, Mg = 40670 and $\Delta t$ = 5 s*

Moreover, at higher temperature gradients and for similar bubble sizes, the same extra downward peak expansions in the transverse direction were observed; however, the stability of the injected bubble was unfortunately affected by the surrounding chaotic flow behavior. This is clearly observed in figure 4.39, where the bubble vibrates in the transverse direction. The time intervals between the successive images shown in these figures were not the same. By repeating the experiment several times, one can conclude that these unwanted bubble vibrations are not caused by system mechanical instabilities during the experiments but caused by the flow behavior itself. In general, the stability of injected bubbles of diameters less than 10 mm is influenced strongly by the chaotic flow behavior at relatively lower Mg-numbers rather than the case of larger bubble sizes. However, at larger bubbles the phase boundary between the bubble and the liquid became unstable at about Mg-numbers > 80000, where these unwanted bubble vibrations altered the bubble volume in a range of ± 5 %.

It was found that, at higher temperature gradients, these non-periodic oscillations started to appear at smaller bubbles rather than at lower temperature gradients. Also, at relatively high temperature gradients the oscillations were observed to act like pulses or waves moving transversely from one side to the other as shown in figure 4.40. These patterns repeated themselves in a non-periodic manner.

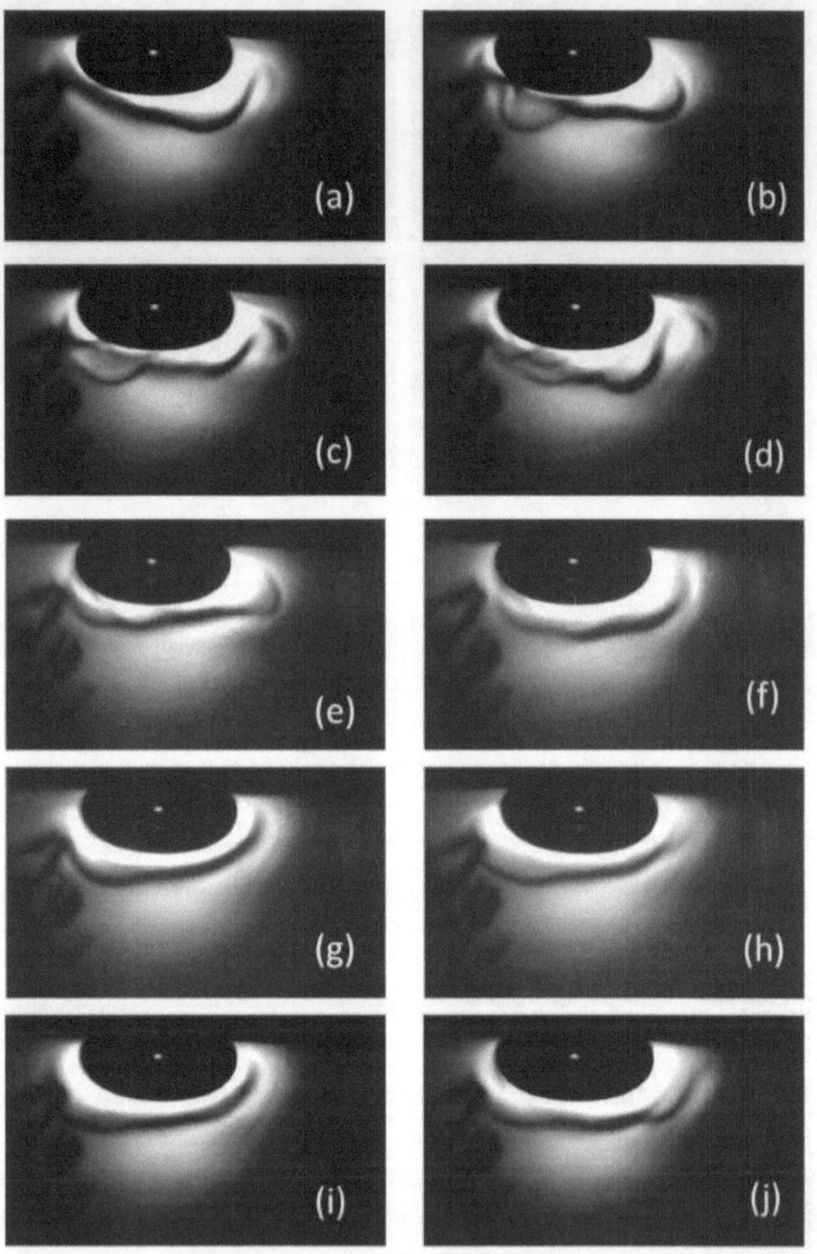

Figure 4.39: Non-Periodic movements of vortex boundary contour, $r_B = 3.77\ mm$, $z_B = 2.55\ mm$, $\frac{r_B}{z_B} = 1.48$, $\Delta p = 0$ bar, $\left|\frac{\partial T}{\partial z}\right| = 2.65\ K/mm$, $Mg = 37775$ and $\Delta t = 2\ s$

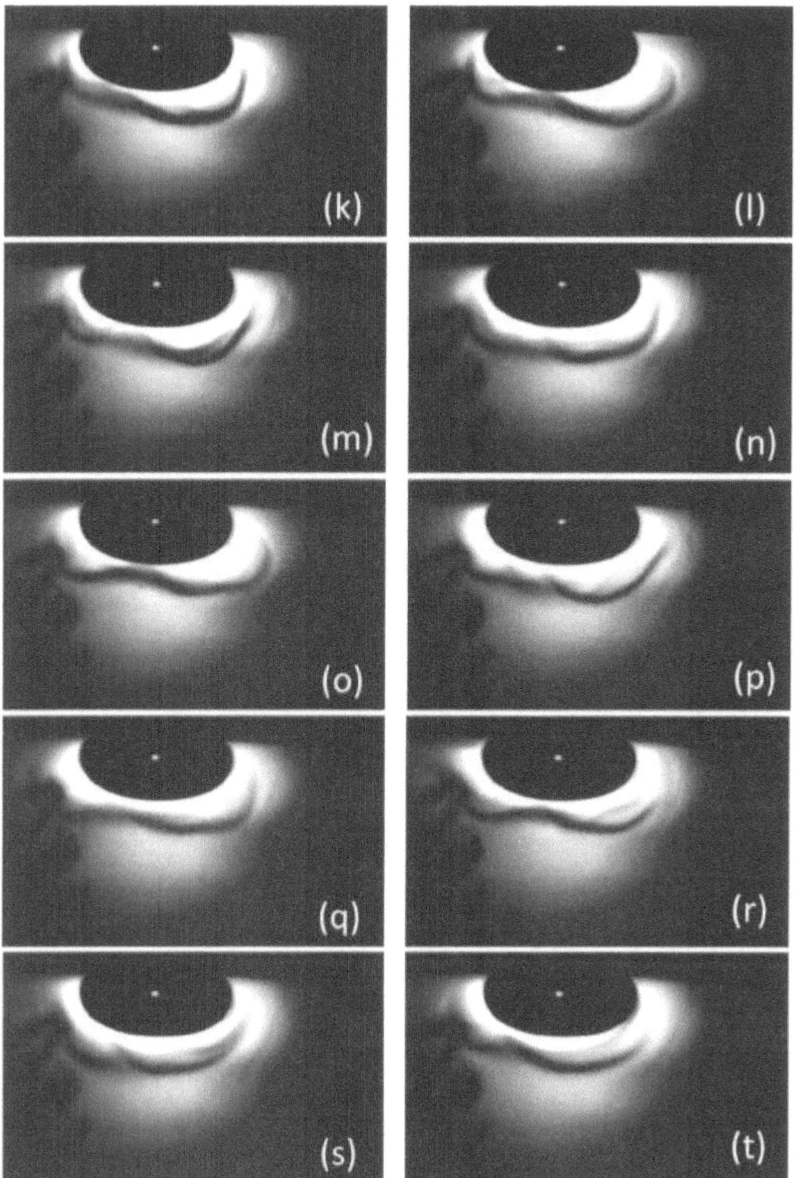

*Figure 4.39: Non-Periodic movements of vortex boundary contour, $r_B = 3.77$ mm, $z_B = 2.55$ mm, $\frac{r_B}{z_B} = 1.48$, $\Delta p = 0$ bar, $\left|\frac{\partial T}{\partial z}\right| = 2.65$ K/mm, Mg = 37775 and $\Delta t = 2$ s*

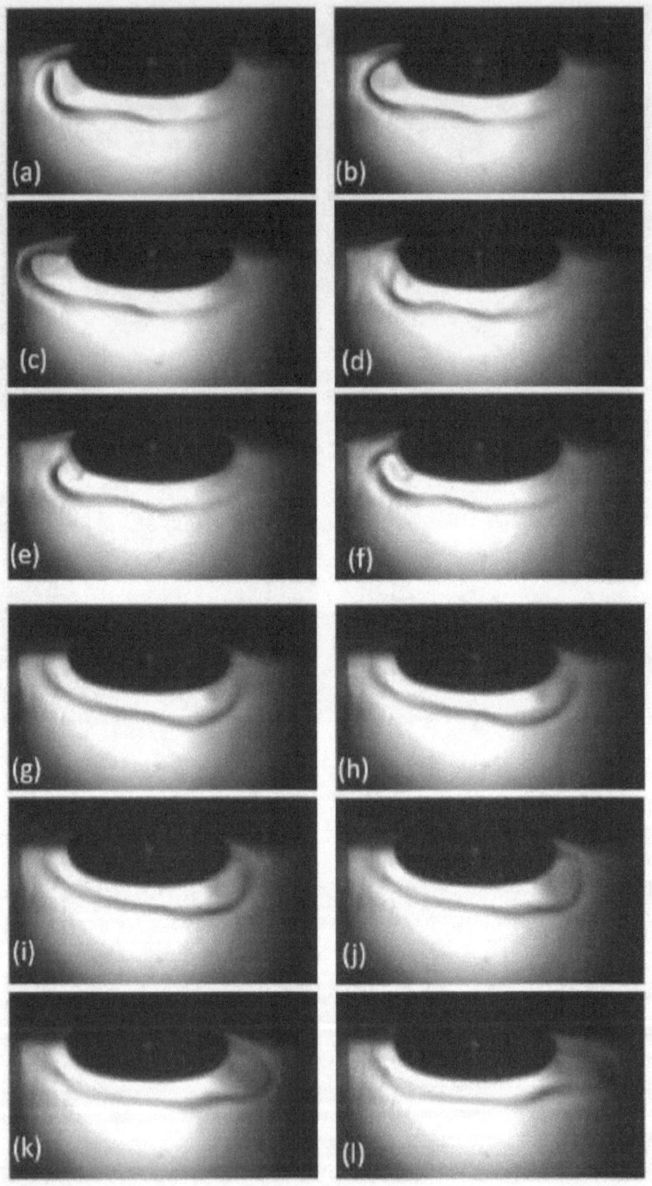

Figure 4.40: Non-Periodic movements of vortex boundary contour, $r_B = 4.75$ mm, $z_B = 2.69$ mm, $\frac{r_B}{z_B} = 1.77$, $\Delta p = 0$ bar, $\left|\frac{\partial T}{\partial z}\right| = 2.63$ K/mm, $Mg = 49870$ and $\Delta t = 3$ s, (a-f: left side oscillations; g-l: right side oscillations)

At smaller bubble sizes ($\frac{r_B}{z_B} < 1.8$) the non-periodic behaviour of the thermocapillary vortex boundary contour oscillations can be evaluated as follows. Figure 4.41 shows that at Mg-numbers higher than the critical one, the boundary contour behaves in that form; it approaches to the bubble from one side while oscillating from the other side in the transverse direction and then it alternates itself (a-d). Another boundary dark contour was also sometimes observed, it cuts the main thermocapillary vortex boundary contour area into two parts (e-h). These kinds of oscillations were also accompanied by vibrations in the bubble itself as mentioned before.

At a relatively higher Mg-number of about 67500, the boundary contour oscillated rapidly and forms a downward peak extension which changed its position along the transverse direction as shown in figure 4.42. Furthermore, figure 4.43 shows internal disorders and disturbances inside the main boundary vortex area in addition to the oscillations of its boundary contour at higher Mg-numbers of about 93000.

At Mg-number of about 101000, a higher degree of oscillations was observed, where many different sized simultaneous eddies take place around the injected air bubble as shown in figure 4.44. The value of the deflection of the central eddy extension is considered to increase with increasing the Mg-number as shown in figure 4.44 (a-d). The presence of such simultaneous eddies (peak extensions) or boundary contour vortices is considered as an obvious indication of turbulence.

Consequently, at high temperature gradients but at larger bubbles, the behaviour of the turbulent boundary contour oscillations develops to a form of a string oscillation mode. The boundary contour acts as a source of hydrothermal waves originating simultaneously from both sides of the injected bubble towards the lateral direction as shown in figure 4.45.

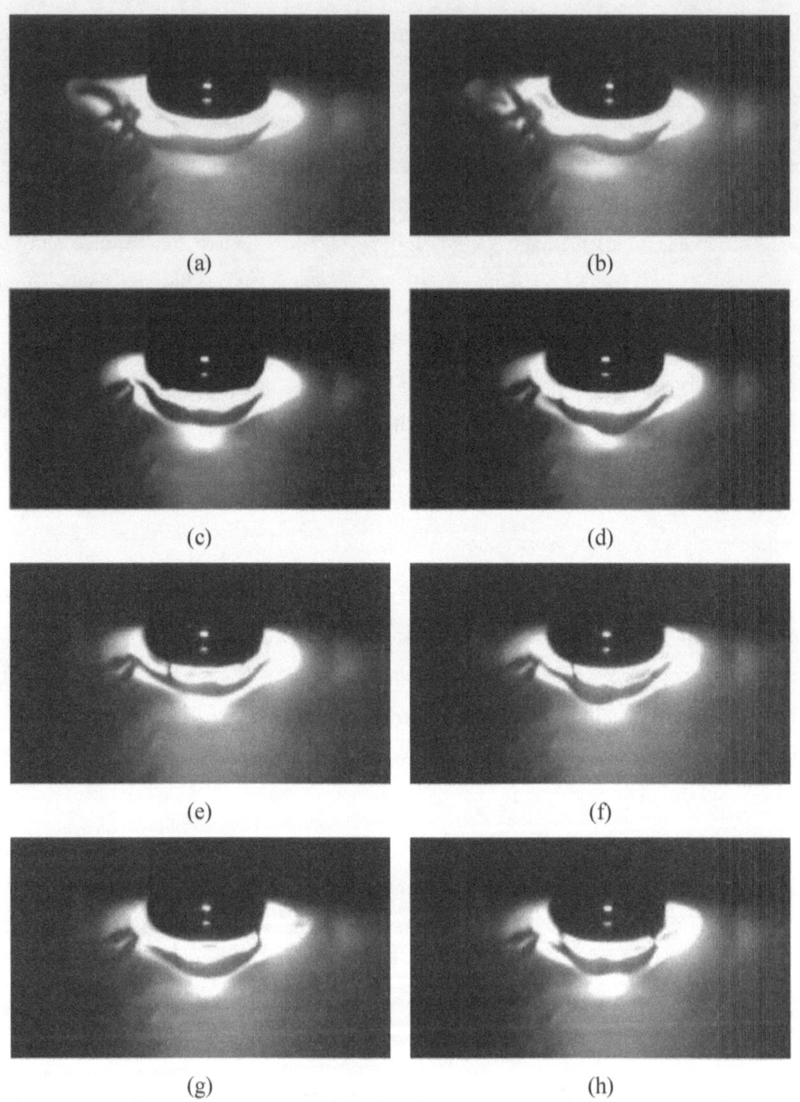

Figure 4.41: Non-periodic movements of the vortex boundary contour, $r_B = 4.2$ mm, $z_B = 2.64$ mm, $\frac{r_B}{z_B} = 1.59$, $\Delta p = 1.9$ bar, $\left|\frac{\partial T}{\partial z}\right| = 3.7$ K/mm, Mg = 65785 and $\Delta t = 0.85$ s

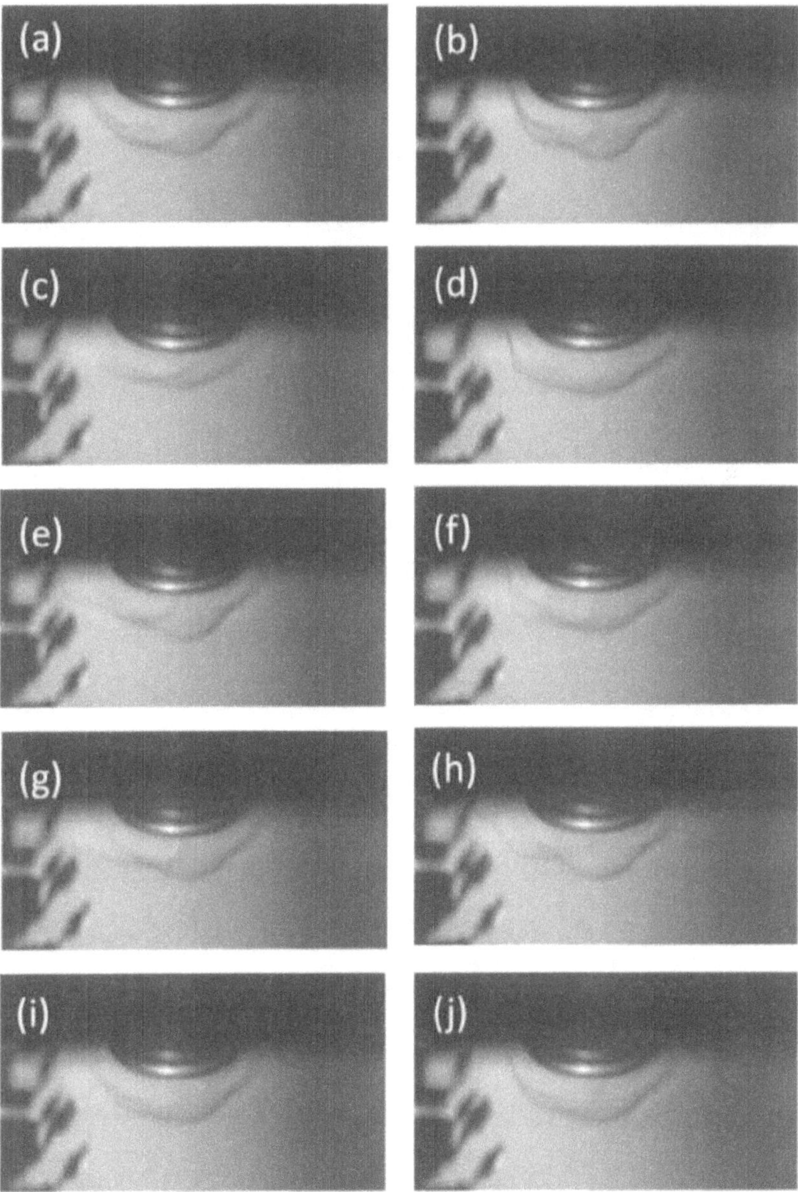

Figure 4.42: Non-Periodic movements of the vortex boundary contour, $r_B = 4\ mm$, $z_B = 2.6\ mm$, $\frac{r_B}{z_B} = 1.54$, $\Delta p = 1.3$ bar, $\left|\frac{\partial T}{\partial z}\right| = 4\ K/mm$, $Mg = 67465$ and $\Delta t = 5\ s$

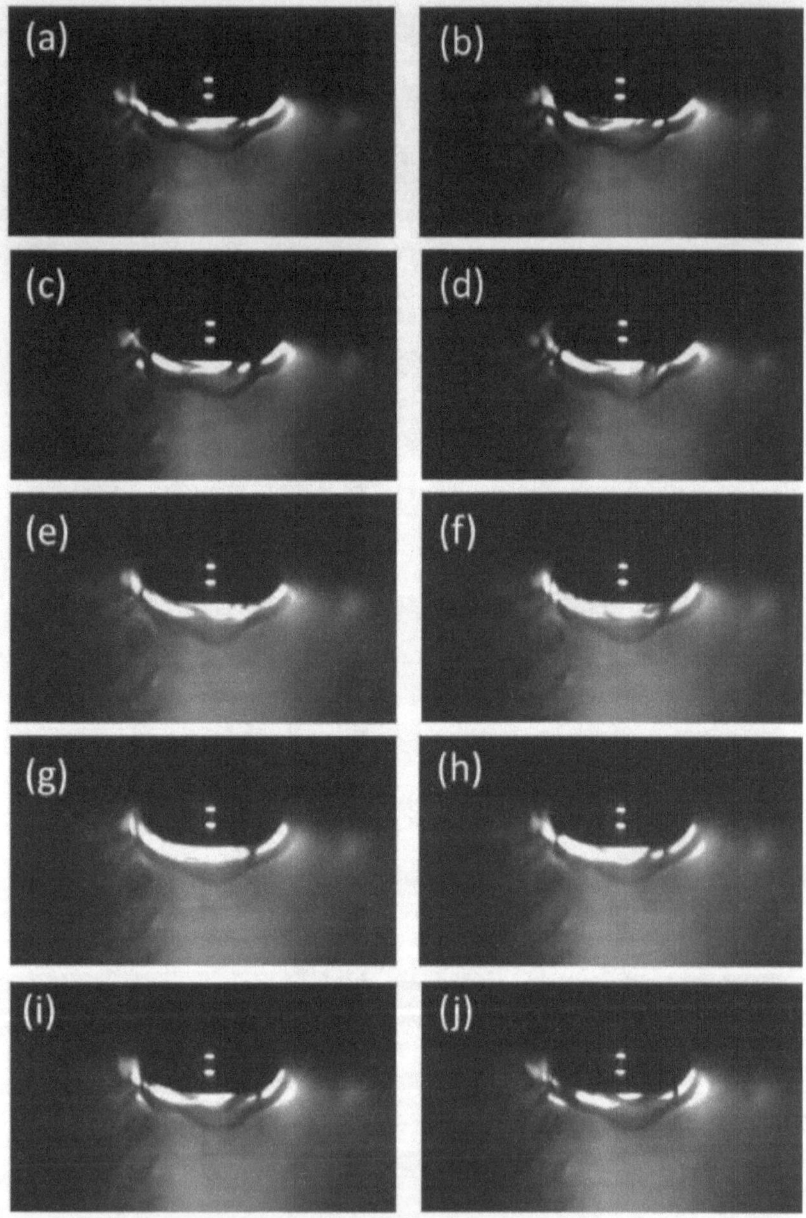

Figure 4.43: Non-periodic movements of the vortex boundary contour, $r_B = 4.85$ mm, $z_B = 2.7$ mm, $\frac{r_B}{z_B} = 1.8$, $\Delta p = 1.9$ bar, $\left|\frac{\partial T}{\partial z}\right| = 4.27$ K/mm, $Mg = 93180$ and $\Delta t = 1.25$ s

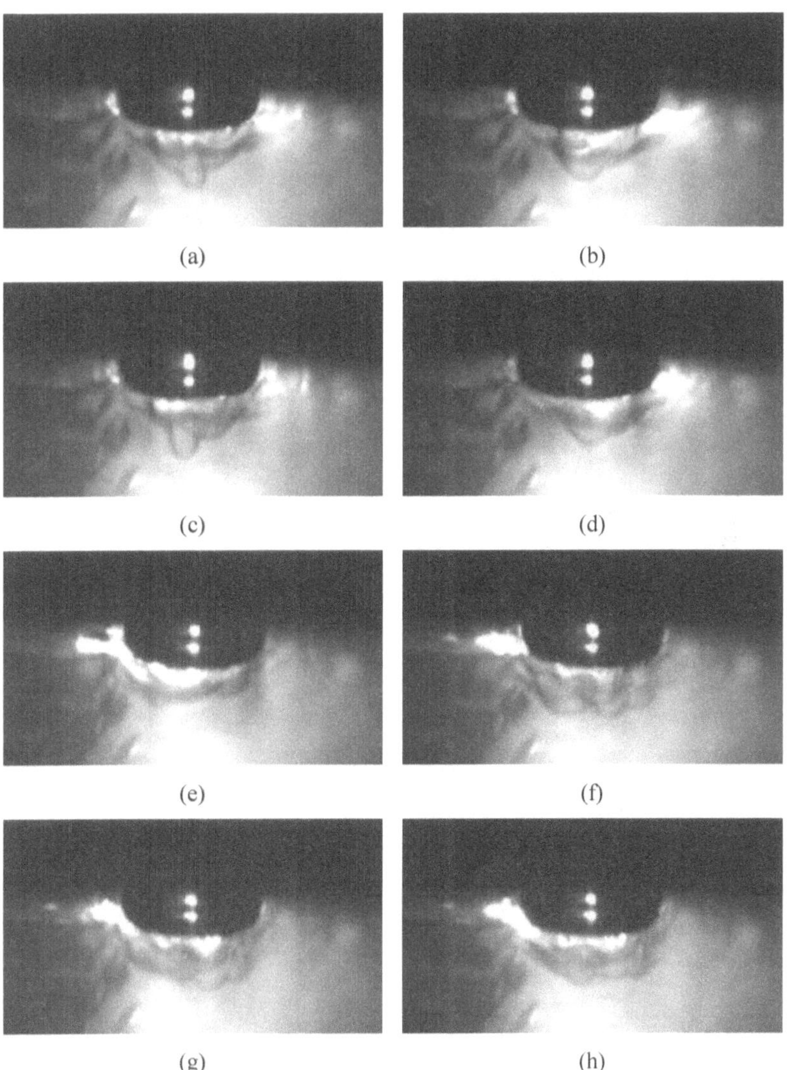

*Figure 4.44: Turbulent behaviour of the vortex boundary contour, $r_B$ = 4.41 mm, $z_B$ = 2.68 mm, $\frac{r_B}{z_B}$ = 1.65, $\Delta p$ = 1.9 bar, $\left|\frac{\partial T}{\partial z}\right|$ = 4.96 K/mm, Mg = 101600 and $\Delta t$ = 0.5 s*

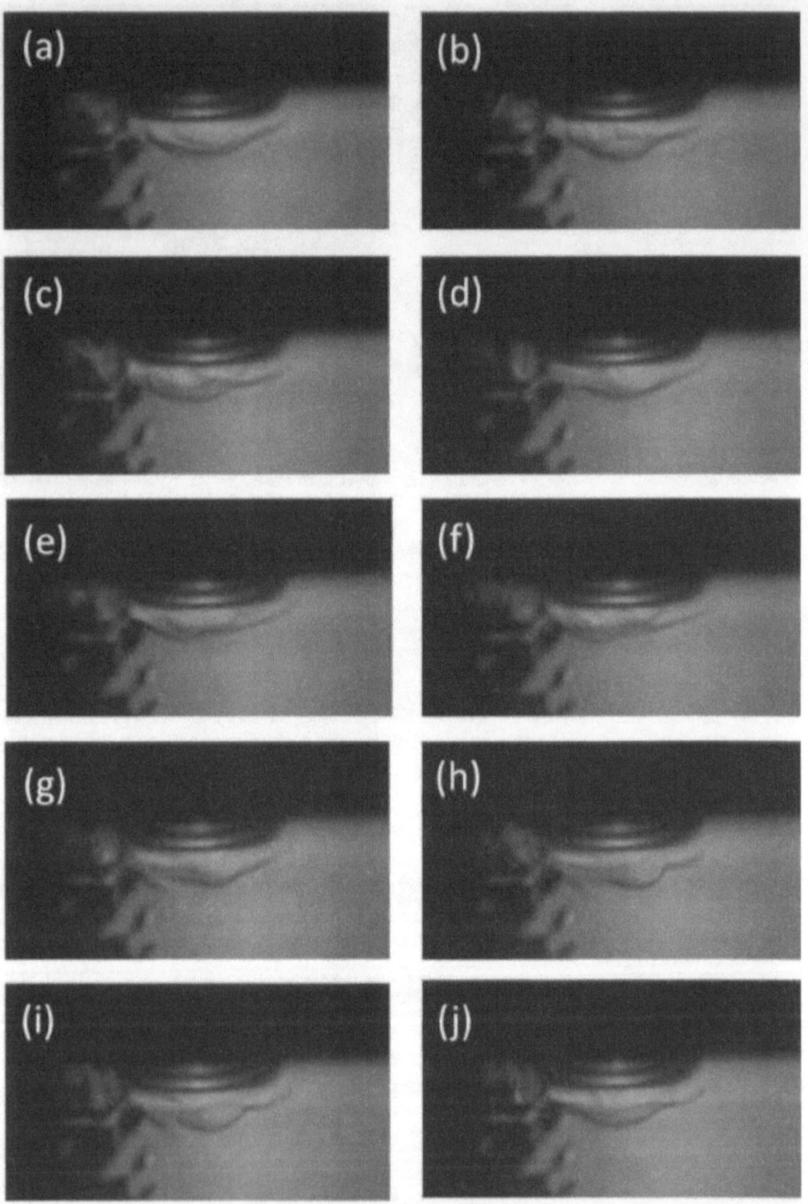

Figure 4.45: Turbulent movements of the vortex boundary contour, $r_B = 5.6$ mm, $z_B = 2.7$ mm, $\frac{r_B}{z_B} = 2.06$, $\Delta p = 2$ bar, $\left|\frac{\partial T}{\partial z}\right| = 5.21$ K/mm, $Mg = 139600$ and $\Delta t = 0.62$ s

### 4.2.3 Classification of The Non-Periodic Oscillatory Thermocapillary Flow

Although various cases of non-periodic oscillations at different bubble sizes were presented before (see figure 4.37 - figure 4.45), the effect of the dimensionless number $\frac{r_B}{z_B}$ on the behaviour of the thermocapillary flow within the non-periodic zone is not well investigated. Therefore, the next few paragraphs demonstrate how that parameter influences the flow behaviour. As mentioned before (see figure 4.24 and figure 4.37), the transitional temperature gradient or Marangoni number is different for every bubble aspect ratio. Small bubbles need higher values of Mg to reach the non-periodic zone of oscillations. Also, within the non-periodic oscillation zone at Mg-numbers more the transitional Marangoni number, the thermocapillary flow undergoes different phases of non-periodic oscillations till reaching a highly developed turbulent state.

For example, for a bubble aspect ratio of 1.61, the corresponding value of the transitional temperature gradient was found to be of about 2 K/mm (Mg = 33400). For an applied temperature gradient of 2.33 K/mm (Mg = 38900), the thermocapillary flow around that bubble was characterized by side oscillations in the vertical direction. These oscillations were similar to the mode of periodic vertical oscillation (shown before in figure 4.21) except that the interactions between the thermocapillary vortex boundary contour itself and the bubble interface formed a central peak of the boundary contour as shown in figure 4.46. This behaviour of non-periodic oscillations we termed **phase A**. At a slightly higher value of the temperature gradient of 2.65 K/mm (Mg = 44300), the fluid flow around the bubble, termed **phase B**, displays a sudden side deformation leading to several central peaks of the boundary contour as shown in figure 4.47. When further increasing the applied temperature gradient to a value of 3.7 K/mm (Mg = 66800), these central peaks were modified and accompanied by fast successive transversely generated side pulses as shown in figure 4.48, where both waves at the left and the right side are displayed. This behaviour of non-periodic oscillations we termed **phase C**. Finally, at a temperature gradient more than 4.9 K/mm (Mg = 99560), the fluid flow around the bubble undergoes **phase D**, which is characterized by a combination between fast side pulses in horizontal direction and central peaks in the vertical direction. This phase is considered as a turbulent state and shown in figure 4.49 and figure 4.50.

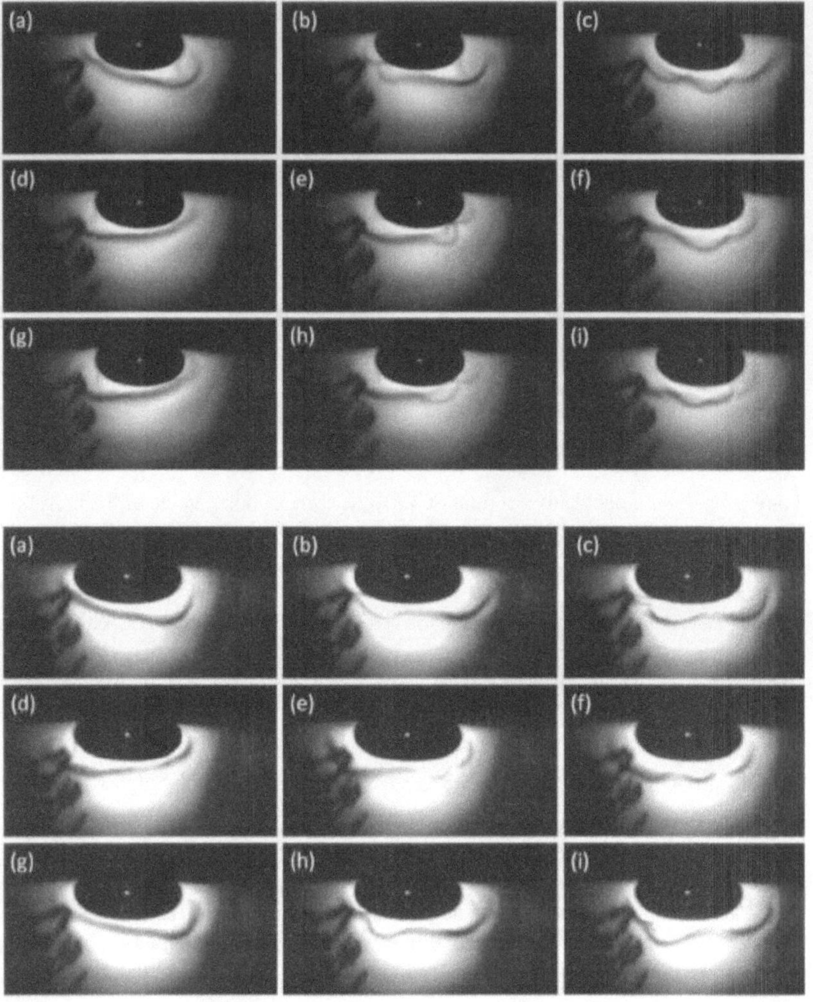

*Figure 4.46: Phase A shadowgraph images, upper: $\frac{r_B}{z_B}$ = 1.42, Mg = 30000 and $\Delta t$ = 2.5 s. Lower: $\frac{r_B}{z_B}$ = 1.61, Mg = 38900 and $\Delta t$ = 2.5 s*

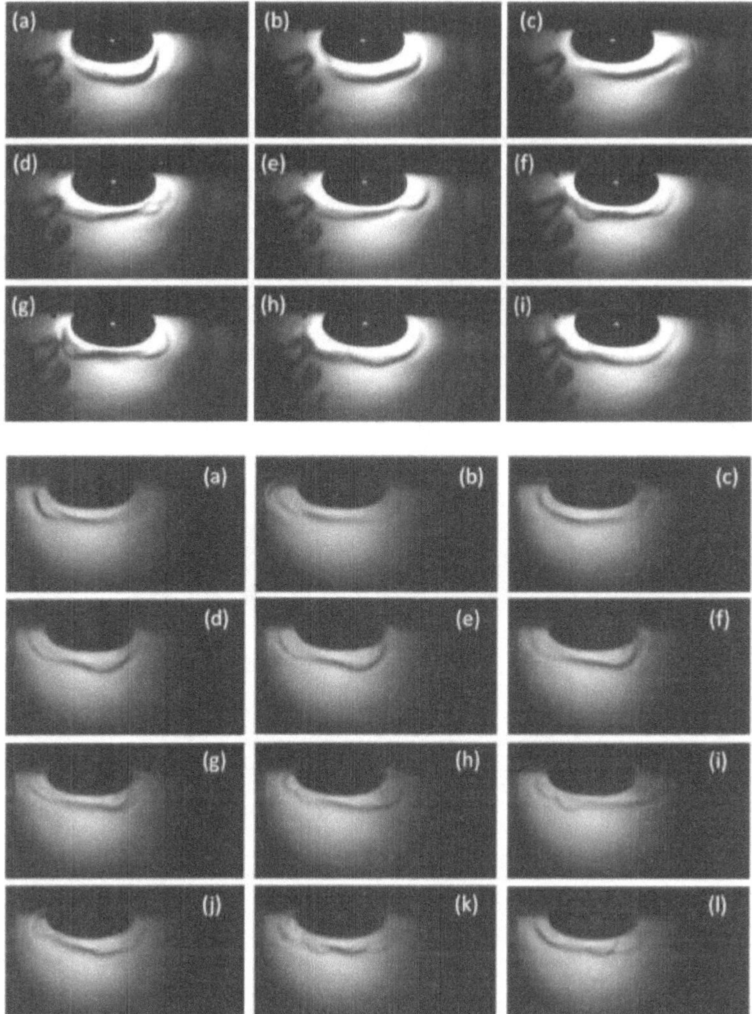

*Figure 4.47: Phase B shadowgraph images, upper: $\frac{r_B}{z_B} = 1.42$, Mg = 36385 and $\Delta t$ = 2.5 s. Lower: $\frac{r_B}{z_B} = 1.61$, Mg = 44300 and $\Delta t$ = 2.5 s*

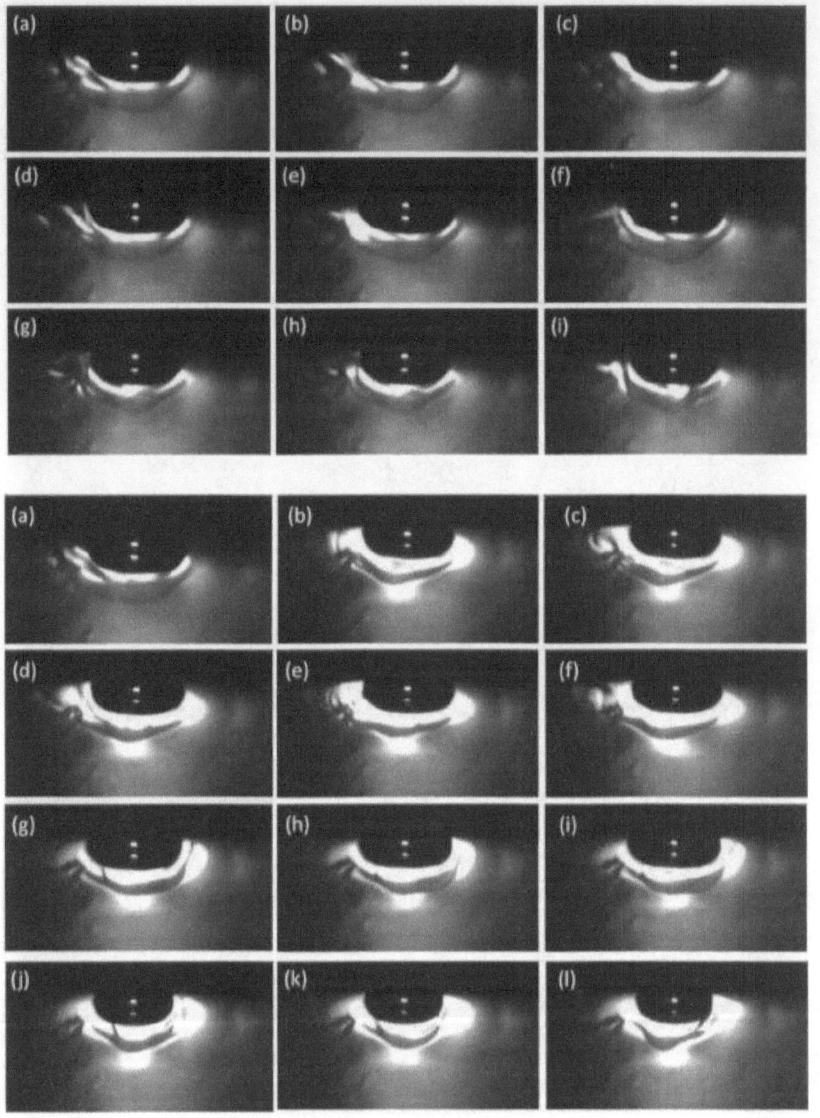

*Figure 4.48: Phase C shadowgraph images, upper: $\frac{r_B}{z_B} = 1.42$, Mg = 69450 and $\Delta t = 0.85$ s.  Lower: $\frac{r_B}{z_B} = 1.61$, Mg = 66800 and $\Delta t = 0.85$ s*

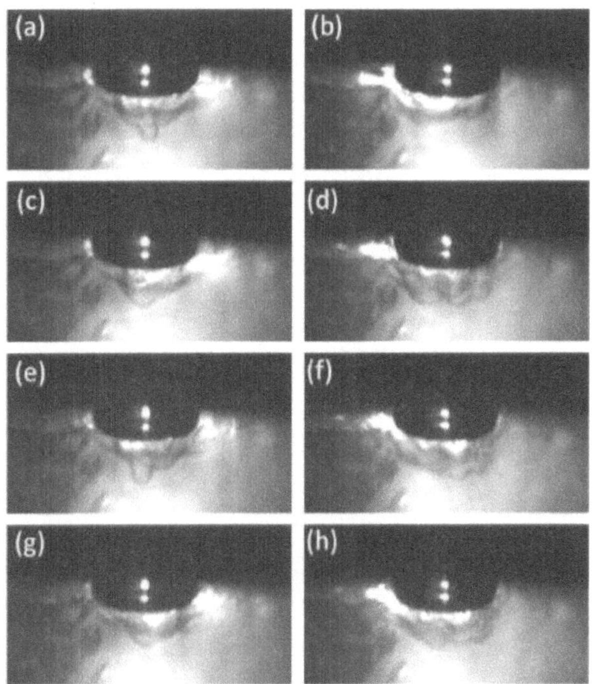

*Figure 4.49: Phase D (turbulence) shadowgraph images, $\frac{r_B}{z_B} = 1.61$, Mg = 99560 and $\Delta t = 0.5$ s*

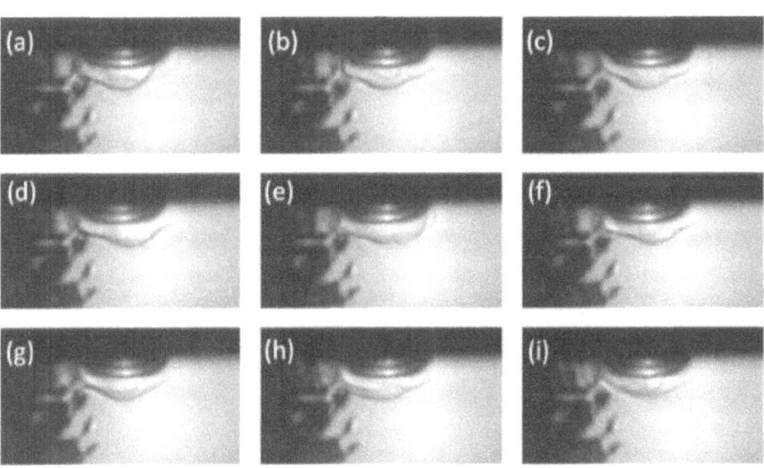

*Figure 4.50: Phase D (turbulence) shadowgraph images, $\frac{r_B}{z_B} = 1.55$, Mg = 97000 and $\Delta t = 0.5$ s*

Figure 4.51 shows comparable temperature plots of different flow modes when increasing Mg-numbers for a bubble aspect ratio of 1.67. Temperature plot (a) reveals a pure periodic fluid flow behaviour within the periodic regime. After that, the non-periodicity starts to appear at a slightly higher Mg- number. Further increase in the Mg-number increases the span of temperature variations and the disorder of temperature fluctuations.

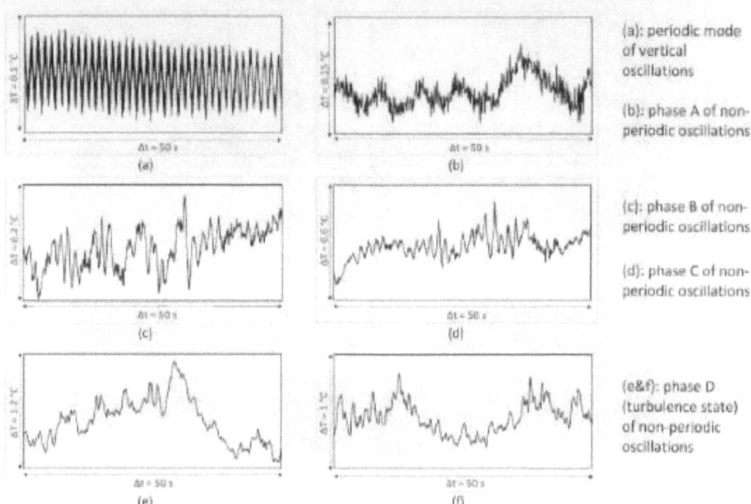


*Figure 4.51: Typical temperature fluctuation plots at point B in the liquid, $r_B$ = 4.5 mm, $z_B$ = 2.7 mm, $\frac{r_B}{z_B}$ = 1.67, (a): Mg = 35450, $\left|\frac{\partial T}{\partial z}\right|$ = 1.96 K/mm, $\Delta P$ = 0 bar, f = 0.67 Hz. (b): Mg = 36400, $\left|\frac{\partial T}{\partial z}\right|$ = 2.04 K/mm, $\Delta p$ = 0 bar. (c): Mg = 62800, $\left|\frac{\partial T}{\partial z}\right|$ = 3.23 K/mm, $\Delta p$ = 1.5 bar. (d): Mg = 85050, $\left|\frac{\partial T}{\partial z}\right|$ = 4.15 K/mm, $\Delta p$ = 1.5 bar. (e): Mg = 135400, $\left|\frac{\partial T}{\partial z}\right|$ = 6.17 K/mm, $\Delta p$ = 1.5 bar. (f): Mg = 139400, $\left|\frac{\partial T}{\partial z}\right|$ = 6.37 K/mm, $\Delta p$ = 1.5 bar*

In addition to that, using statistical methods, the data of the temperature fluctuation plots were analyzed to provide a more quantitative characterization of the different identified convective flow modes. The temperature field around the air bubble is affected by the thermocapillary fluid flow and thus the flow velocities and their fluctuations. Hence, we quantify the intensity of the temperature fluctuations of the flow modes using the quantity T (fluid temperature). The mean temperature is defined as

$$\bar{T} = \frac{1}{n}\sum_{1}^{n} T_i .$$
(4.1)

Using $T_i' = T_i - \bar{T}$
(4.2)

and $T_{rms}' = \sqrt{\frac{1}{n}\sum_{1}^{n}(T_i')^2}$ ,
(4.3)

the turbulence intensity Tu is defined as $Tu = \frac{T_{rms}}{\bar{T}} \times 100\ \%$. $\hspace{2cm}$ (4.4)

$\bar{T}$ denotes the average value of n discrete temperature measurements ($T_i$), $T_i'$ the temperature fluctuations, $T_{rms}'$ the root-mean-square value of temperature fluctuations and Tu the turbulence intensity (magnitude of turbulence).

The temperature oscillations, shown in figure 4.51 (a), show a maximum value $T_{max}'$ of about 0.04 K and a Tu of 0.061 %. That reveals a periodic thermocapillary flow behaviour, which is noticeable in the corresponding temperature fluctuation plot. Figure 4.51 (b) shows the start of the flows non-periodicity, where the values of $T_{max}'$ and Tu increase to 0.063 K and 0.084 %, respectively. Further increasing Mg-numbers leads to upgrading the non-periodic phase of fluctuations to higher modes. The temperature fluctuations of phase B, shown in figure 4.51 (c), show a maximum value $T_{max}'$ of about 0.09 K and a Tu of 0.09 %. Furthermore, The temperature fluctuations of figure 4.51 (d), which corresponds to phase C of non-periodic oscillations, show higher values of $T_{max}'$ and Tu of 0.18 K and 0.19 %, respectively. Henceforth, the values of $T_{max}'$ and Tu increase till reaching about 0.22 K and 0.32 %, respectively, see temperature fluctuation plots of figure 4.51 (e & f), which reveal a noticeable turbulence intensity of a developed turbulent thermocapillary flow.

A temperature gradient of 4.96 K/mm was sufficient for a bubble of $\frac{r_B}{z_B} = 1.62$ to reach the turbulent phase D, however a smaller bubble of $\frac{r_B}{z_B} = 1.55$ needed a higher temperature gradient of 5.21 K/mm to reach that phase. Thus, the critical values of temperature gradients and accordingly Mg-numbers, needed to upgrade a non-periodic phase of oscillation to the next higher one, depend on the temperature gradient and the bubble aspect ratio, which determines the shape (curvature) of the phase boundary (interface) between the two different phases (liquid and gas). It was noticed also that for moderate bubble sizes of bubble aspect ratios between 1.5 to 1.65, a value of Mg-number of order of 95000 is enough to reach the turbulent state as shown in figure 4.52. This figure displays a summary of our experimental results in form of a Mg, $(\frac{r_B}{z_B})$-map. It shows the boundary between the periodic flow conditions and the non-periodic ones. All experiments are comparable regarding fluids (silicone oil), dimension of the test cell and boundary conditions. We only varied temperature gradients and bubble sizes. Also, it indicates the boundary between periodic flow conditions and the non-periodic ones. Moreover, it provides guideline borders between the periodic and the non-periodic flow mode regimes (phase A-D). The measuring points were generated under variation of the bubble aspect ratio in a range of 1.2 to 2.4 and the temperature gradient $\left|\frac{\partial T}{\partial z}\right|$ up to 6.5 K/mm (Mg between 10000 and 140000).

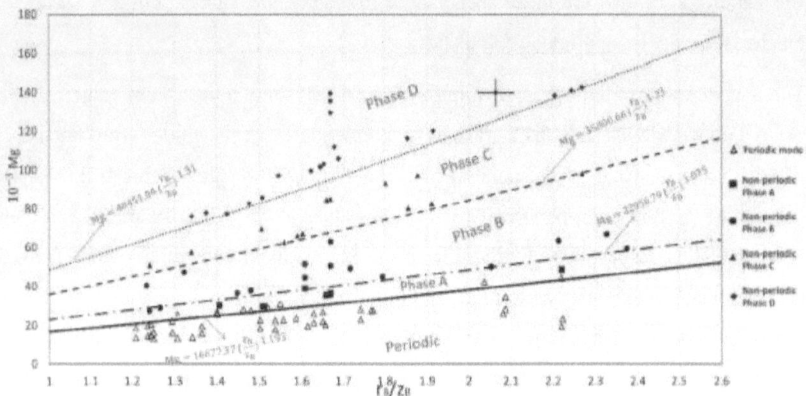



*Figure 4.52: Mg, $\left(\frac{r_B}{z_B}\right)$ map: summary of measurements and classification of flow modes*

As the measurement data of figure 4.52 suggest, a simple power law of the form

$$Mg_{i \rightarrow j} = A \left(\frac{r_B}{z_B}\right)^n \tag{4.5}$$

can be derived within the range of the measurement data in order to express the phase transient border lines in form of empirical equations. The values of the corresponding constants A and n are given in table 3.

| $i \rightarrow j$ | A | n | Symbol |
|---|---|---|---|
| *Periodic → Phase A* | 16677.37 | 1.193 | ———————— |
| *Phase A → Phase B* | 22956.79 | 1.075 | — · — · — |
| *Phase B → Phase C* | 35800.66 | 1.23 | — — — — — · |
| *Phase C → Phase D* | 48451.04 | 1.31 | ················· |

*Table 3: Values of power-law constants of the phase transient border lines*

The trends of the power-law functions show, that the first non-periodic mode (phase A) takes place in a relatively narrow range of Mg versus $\frac{r_B}{z_B}$. The sectors of phases B, C and D are rather wide, and their corresponding boundaries increase with increasing the exponent n the value of which is given in table 3. It should also be noted, that for physical reasons these transient border lines have to go through the origin of the plot Mg versus $\frac{r_B}{z_B}$, figure 4.52.

# 5 CONCLUSIONS AND RECOMMENDATIONS

## 5.1 Conclusions

We experimentally studied both the periodic and non-periodic oscillatory flow of the thermocapillary convection around an air bubble under a heated wall under gravitational conditions, using a low viscosity liquid as working fluid. Shadowgraph and PIV techniques were applied to visualize the movement of the thermocapillary vortex boundary contour and measure the fluid velocity, respectively. Moreover, local temperature measurements delivered typical temperature fluctuation plots of the identified flow modes. We have been able to reach high Mg-numbers by conducting the experiments under excess pressure conditions. The study reveals the origin of the boundary contour turbulent oscillations at high Mg-numbers and shows how the fluid motion develops from a periodic oscillatory state to a non-periodic state in dependence of the Mg-number and the bubble aspect ratio. The PIV velocity vector field images show that the thermocapillary boundary contour oscillations are a result of the flow vortex movements and the variation of its size. We have observed three modes of periodic oscillations: (i) vertical oscillations, (ii) transverse symmetric oscillations, and (iii) transverse asymmetric oscillations and have shown the vertical periodic oscillatory state through both transient temperature and velocity measurements. Moreover, non-periodic oscillations of the boundary contour were achieved at higher Mg-numbers. The transitional temperature gradients respectively the $Mg_{tran}$-numbers changed with the bubble aspect ratio. In other words, the temperature gradient and accordingly the Mg number required to shift the flow from one phase to another depends on the bubble aspect ratio. That confirms that the bubble aspect ratio is a very determined geometrical parameter to indicate the thermocapillary bubble convection behaviour. That demonstrates that, in other configurations, the shape of the interface between the two fluids in any surface tension driven flow configurations, is very important to determine the flow behaviour and its development at very high Mg-numbers.

Additionally, our results using horizontal plane visualization demonstrated that the thermocapillary convective flow within the periodic mode of oscillations is characterized by fluid oscillations in the horizontal plane (parallel to the heated wall) as well.

Our experiments reveal that in a large range of bubble sizes, the non-periodic oscillation state of the fluid flow around the gas bubble undergoes four different modes (A-D) in dependence of Mg. The last one (phase D) is a developed turbulent state starting at Mg-numbers of 75000 for the smallest bubble aspect ratio $\frac{r_B}{z_B} = 1.2$ up to the maximum measured Mg-number = 140000 for a bubble aspect ratio $\frac{r_B}{z_B} = 2.3$. These different phases were distinguished through various modes of the boundary contour fluctuations. Different behaviours of temperature fluctuation plots for the thermocapillary flow modes are briefly shown in figure 5.1.

In addition, thermocapillary flow driving velocities were measured at different boundary conditions (bubble sizes) close to the bubble surface. The measurements demonstrated that the magnitude of the thermocapillary fluid flow velocity at high Mg can reach a value of the order of 3 cm/s, which equals around 75 times the mean velocities of the thermocapillary convective steady laminar fluid flow. The data clearly reveal that it is the high magnitude of the interfacial velocity that initiates the interactions between thermocapillary flow vortices leading finally to the non-periodic boundary contour oscillations and hence buoyancy plays a secondary role in the described flow configuration. Eventually, we could satisfactorily proof turbulent thermocapillary flow.

Many questions remain. It would be interesting to look at smaller bubbles of aspect ratios between 0 and 1, which become more spherical with decreasing size. This change of surface curvature would markedly influence the flow and temperature field. However, corresponding experiments within this aspect ratio range necessitate a completely newly reviewed measurement technique for the need of far higher image resolutions. Also, continuative experiments under low-gravity conditions would allow to decouple buoyant convection from the surface tension driven one. We expect a far wider flow field due to thermocapillary convection under weightless conditions, as already shown for laminar convection at very low Marangoni numbers in a sounding rocket experiment [34]. Another important question is, how the instant surface temperature gradient at the bubble periphery develops when increasing Mg. Experimentally, this information is extremely difficult to catch. Here, numerical simulations could give an answer. However, resolution of the remaining problems will have to await further studies on the experimental as well on the theoretical/numerical side.

Finally, both velocity and temperature fields of the thermocapillary flow around the bubble are influenced in all directions where no symmetricity could be attained to simplify the flow configuration into a 2-d model. However, through temperature measurements, 2-d velocity field measurements and 2-d shadowgraphy, the current study illuminated well the thermocapillary bubble convection and found a systematic way to categorize the fluid flow regarding various modes occurring beyond the Mg$_{tran}$ through the behaviour of the thermocapillary vortex boundary contour oscillations.

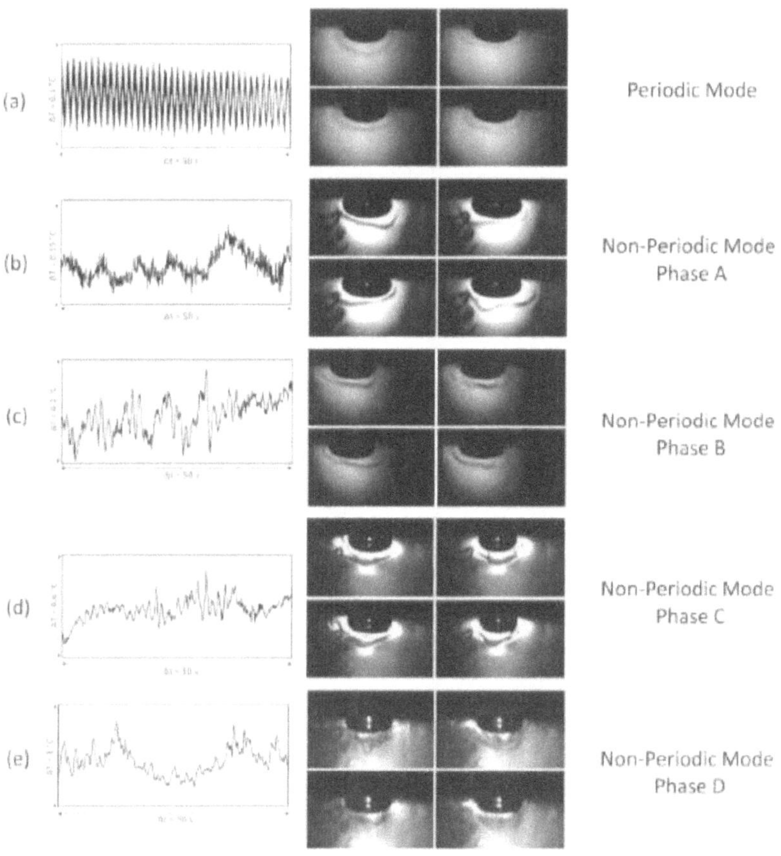

*Figure 5.1: Typical temperature fluctuation plots and shadowgraph images of thermocapillary bubble convection modes*

## 5.2 Recommendations for Future Work

As mentioned in the current study and the literature as well, the gravitational force has an influence on the thermocapillary bubble flow, where it affects the shape of the flow vortices and the shape of the liquid-air interface. Therefore, it is suggested to conduct these experiments under gravitationless conditions to eliminate the influence of buoyancy.

In the current study the turbulent thermocapillary bubble convection is explored, however, its effect on the contact angle between the bubble periphery and the upper heated wall needs more detailed illustration. Therefore, it is recommended to use the same experimental setup (shadowgraphy) to define an experimentally concluded correlation describing the relation between the contact angle and the Mg-number.

Liquid properties change dramatically at high temperatures; therefore, the phase interface (bubble geometrical shape) is affected by temperature gradients particularly at high Marangoni numbers, which needs a more detailed understanding.

In addition to that the applied temperature gradient in z direction in the current study is estimated through temperature sensors installed inside the test cell. Also, a better estimation of the effective temperature gradient on the bubble interface within the very dynamic Marangoni convection could help in a more accurate definition of the Mg-number. Also, we are fully aware of the effect that this value is modulated at the interface due to the different curvatures of the applied bubbles and the different flow states as well. The question of the real existing temperature distribution on the surface is extremely difficult to gain experimentally, therefore, it is recommended to pursue numerical simulations to predict the real value of the temperature gradient at the phase interface.

Regarding the set of experiments done in the current study, it is recommended to pursue the same experiments and reach the four modes of non-periodic thermocapillary oscillations at larger bubble aspect ratios of $\frac{r_B}{z_B} > 2.4$ and smaller ones of $\frac{r_B}{z_B} < 1.2$.

Future scientists are able of using the current experimental results for model-validation when developing numerical models to numerically explore the dynamic thermocapillary flow around a gas bubble.

## REFERENCES

[1]   J. Aleksic, J. A. Szymczyk, A. Leder, and T. A. Kowalewski, "Experimental investigations on thermal, thermocapillary and forced convection in Czochralski crystal growth configuration," *CMEM 2001, Computational Methods and Experimental Measurements X*. WIT Press, Alicante, Spain, pp. 627–636, 2001.

[2]   G. Amberg and J. Shiomi, "Thermocapillary flow and phase change in some widespread materials processes," *FDMP-Fluid Dyn. Mater. Process.*, vol. 1, no. 1, pp. 81–96, 2005.

[3]   R. S. Subramanian, R. Balasubramaniam, and G. Wozniak, "Fluid mechanics of bubbles and drops," in *Physics of Fluids in Microgravity*, R. Monti, Ed. Taylor and Francis, 2002, pp. 149–177.

[4]   S. H. Jin *et al.*, "Using nanoscale thermocapillary flows to create arrays of purely semiconducting single-walled carbon nanotubes," *Nature Nanotech.*, vol. 8, pp. 347–355, 2013.

[5]   J. Cai, F. Liu, and W. A. Sirignano, "Opposed flow impact on flame spread above liquid fuel pools," in *Proc. Western States Sect. Comb. Inst. (WSS/CI), San Diego, CA, USA*, 2002, pp. 1–20.

[6]   J. Thomson, "On certain curious motions observable at the surface of wine and other alcoholic liquors," *Philos. Mag.*, vol. 10, pp. 330–333, 1855.

[7]   C. G. M. Marangoni, "Ueber die Ausbreitung der Tropfen einer Fluessigkeit auf der Oberflaeche einer anderen," *Ann. Rev. Phys. Chem*, vol. 7, pp. 337–354, 1871.

[8]   H. Bénard, "Les tourbillons cellulaires dans une nappe liquide. Premiere partie: description generale des phenomenes," *Rev. Gen. Sci. Pures Appl.*, vol. 11, pp. 1261–1271, 1900.

[9]   L. Rayleigh, "On convection currents in a horizontal layer of fluid, when the higher temperature is on the under side," *Philos. Mag.*, vol. 32, no. 192, pp. 529–546, 1916.

[10]   J. R. A. Pearson, "On convection cells induced by surface tension," *J. Fluid Mech.*, vol. 4, no. 5, pp. 489–500, 1958.

[11] A. Nepomnyashchy, M. G. Velarde, and P. Colinet, "Thermocapillary and solutocapillary migration of drops (and bubbles) and their spreading due to the Marangoni effect," in *Interfacial phenomena and convection*, First., Boca Raton: Chapman & Hall/CRC, 2002, pp. 39–90.

[12] E. S. Gaddis, "The thermal equilibrium of a vapour bubble on a heated solid surface (Ph.D. book)," Univ. Manchester, Manchester, England, 1968.

[13] J. L. McGrew, F. L. Bamford, and T. R. Rehm, "Marangoni flow: an additional mechanism in boiling heat transfer," *Science*, vol. 153, no. 3740, pp. 1106–1107, 1966.

[14] Y. S. Kao and D. B. R. Kenning, "Thermocapillary flow near a hemispherical bubble on a heated wall.," *J. Fluid Mech.*, vol. 53, pp. 715–735, 1972.

[15] B. K. Larkin, "Thermocapillary flow around hemispherical bubble," *AIChe J.*, vol. 16, pp. 101–107, 1970.

[16] V. Huplik and G. D. Raithby, "Surface-tension effects in boiling from a downward-facing surface," *J. Heat Transf.*, vol. 94, pp. 403–409, 1972.

[17] G. Wozniak, J. Siekmann, and J. Srulijes, "Thermocapillary bubble and drop dynamics under reduced gravity - survey and prospects," *Z. Flugwiss. Weltraumforsch. [ZDB]*, vol. 12, pp. 137–144, 1988.

[18] Y. Li, "Experimental Studies of Marangoni Convection With Buoyancy in Simple and Binary Fluids Experimental Studies of Marangoni Convection (Ph.D. book)," Dept. Mech. Eng., Georgia Institute of Technology, Atlanta, Georgia, USA, 2015.

[19] M. F. Schatz and G. P. Neitzel, "Experiments on thermocapillary instabilities," *Annu. Rev. Fluid Mech.*, vol. 33, pp. 93–127, 2001.

[20] J. A. Szymczyk and J. Siekmann, "Experimental investigation of the thermo-capillary flow along cylindrical interfaces under rotation," in *Proc. 7th Symp. on Materials and Fluid Sciences in Microgravity, Oxford, UK*, 1989, pp. 315–320.

[21] R. Velten, D. Schwabe, and A. Scharmann, "Gravity-dependence of the instability of surface-tension-driven flow in floating zones," in *Proc. 7th Symp. on Materials and Fluid Sciences in Microgravity, Oxford, UK*, 1989, pp. 271–278.

[22]  D. Raake, J. Siekmann, and C. H. Chun, "Temperature and velocity fields due to surface tension driven flow," *Exp. Fluids*, vol. 7, pp. 164–172, 1989.

[23]  G. Wozniak and K. Wozniak, "Buoyancy and thermocapillary flow analysis by the combined use of liquid crystals and PIV," *Exp. Fluids*, vol. 17, pp. 141–146, 1994.

[24]  K. Wozniak, G. Wozniak, and T. Rösgen, "Particle-image-velocimetry applied to thermocapillary convection," *Exp. Fluids*, vol. 10, pp. 12–16, 1990.

[25]  C. H. Chun, D. Raake, and G. Hansmann, "Oscillating convection modes in the surroundings of an air bubble under a horizontal heated wall," *Exp. Fluids*, vol. 11, pp. 359–367, 1991.

[26]  C. Reynard, R. Santini, L. Tadrist, and P. Arlabosse, "The experimental study of the periodic instability of thermocapillary convection around an air bubble," in *Proc. STAIF. Space Technology and Applications International Forum, Albuquerque, USA*, 2000.

[27]  C. Reynard, R. Santini, and L. Tadrist, "Experimental study of the gravity influence on the periodic thermocapillary convection around a bubble," *Exp. Fluids*, vol. 31, pp. 440–446, 2001.

[28]  J. Betz and J. Straub, "Numerical and experimental study of the heat transfer and fluid flow by thermocapillary convection around gas bubbles," *Heat Mass Transfer*, vol. 37, pp. 215–227, 2001.

[29]  K.-P. Schade, G. Wozniak, and C. Gessenhard, "Experimental investigation of thermocapillary flow at the phase boundary of a bubble in a liquid," in *Proc. 6th International Conference on Multiphase Flow, Leipzig, Germany*, 2007, no. 902, pp. 1–9.

[30]  G. Christopher, "Aufbau, Inbetriebnahme und Test eines Differential-Interferometers (Diplomarbeit)," Dept. Mech. Eng., Tech. Univ. Chemnitz, Chemnitz, Germany, 2007.

[31]  K.-P. Schade, G. Wozniak, and D. Rubes, "Unstable thermocapillary flow at a gas-liquid phase boundary," in *Proc. Appl. Math. Mech. PAMM*, 2009, vol. 9, pp. 455–456.

[32]  G. Wozniak, "On the thermocapillary motion of droplets under reduced gravity," *J. Colloid Interface Sci.*, vol. 141, no. 1, pp. 245–254, 1991.

[33]  N. 0. Young, J. S. Goldstein, and M. J. Block, "The motion of bubbles in a vertical temperature gradient," *J. Fluid Mech.*, vol. 6, no. 3, pp. 350–356, 1959.

[34]  G. Wozniak, K. Wozniak, and H. Bergelt, "On the influence of buoyancy on the surface tension driven flow around a bubble on a heated wall," *Exp. Fluids*, vol. 21, pp. 181–186, 1996.

[35]  G. Wozniak, K. Wozniak, and J. Siekmann, "Non-isothermal flow diagnostics using microencapsulated cholesteric particles," *Appl. Sci. Res.*, vol. 56, pp. 145–156, 1996.

[36]  R. J. Adrian, "Particle-imaging techniques for experimental fluid mechanics," *Annu. Rev. Fluid Mech*, vol. 23, pp. 261–304, 1991.

[37]  R. J. Keane, R.D., Adrian, "Theory of cross-correlation analysis of PIV images.," *Appl. Sci. Res.*, vol. 49, pp. 191–215, 1992.

[38]  R. J. Adrian, "Dynamic ranges of velocity and spatial resolution of particle image velocimetry," *Meas. Sci. Technol.*, vol. 8, pp. 1393–1398, 1997.

[39]  T. Roesgen, K. Wozniak, and G. Wozniak, "Image processing for laser speckle velocimetry using 2-D fast Fourier transform," *Appl. Opt.*, vol. 29, no. 35, pp. 5298–5302, 1990.

[40]  K. Wozniak and G. Wozniak, "Experimental Analysis of Bénard-Convection by the Combined Application of liquid crystals and PIV," in *Proc. VIIIth European Symposium on Materials and Fluid Sciences in Microgravity*, 1992.

[41]  G. Wozniak and K. Wozniak, "Simultaneous measurement of temperature and velocity fields in thermo-convective liquid flows-Development of a new measuring technique," *Adv. Space Res.*, vol. 11, no. 7, pp. 33–38, 1991.

[42]  H. jr. Oertel, "Thermische Zellularkonvektion, Habilitationsschrift," Univ. Karlsruhe, Karlsruhe, Germany, 1979.

[43]  K. Bühler, "Zellularkonvektion in rotierenden Behältern (Ph.D. book)," Univ. Karlsruhe, Karlsruhe, Germany, 1979.

[44]  J. A. Srulijes, "Zellularkonvektion in Behältern mit horizontalen Temperaturgradienten (Ph.D. book)," Univ. Karlsruhe, Karlsruhe, Germany, 1979.

[45]  K. R. Kirchartz, "Zeitabhängige Zellularkonvektion in horizontalen und geneigten Behältern," Karlsruhe university, 1980.

[46] G. Wozniak, "Optical whole-field methods for thermo-convective flow analysis in microgravity," *Meas. Sci. Technol.*, vol. 10, pp. 878–885, 1999.

[47] K.-P. Schade and G. Wozniak, *Untersuchung oszillatorische und turbulenter Strömungszustände der thermokapillaren Konvektion einer Blase an einer Wand.* Professur Strömungsmechanik, Tech. Univ. Chemnitz, Chemnitz, Germany, Tech. Rep. DFG- WO406/1-2, 2010.

[48] W. Merzkirch, "Optical flow visualization," in *Flow Visualization*, 2nd edn., Academic Press, 1987, pp. 115–231.

[49] H. D. Xi, S. Lam, and K. Q. Xia, "From laminar plumes to organized flows: The onset of large-scale circulation in turbulent thermal convection," *J. Fluid Mech.*, vol. 503, pp. 47–56, 2004.

[50] *Rules of construction of pressure vessels, ASME Boiler and Pressure Vessel Code, ASME BPVC. VIII-1.* 2019.

[51] "Ilmadur® borosilicate gauge sight glasses," *Ilmaborglass GmbH, Ilmenau, Germany*, 2015. [Online]. Available: https://www.glassglobal.com/directory/glass/profile/documents/file.asp?AdrID=83 58&ID=9058.

[52] "MAXOS® Sicherheits-Schaugläser Spezialgehärtet," *Auer Lighting GmbH, Bad Gandersheim, Germany*, 2020. [Online]. Available: https://www.auer-lighting.com/fileadmin/user_upload/auer_MAXOS_Brosch_2020_DE_web.pdf.

# Appendix   DESIGN OF THE PRESSURE CHAMBER

The pressure chamber, described in chapter 3.3 is, in origin, fabricated for other research purposes to withstand vacuum pressure. Therefore, a detailed mechanical analysis of every component of the pressure chamber against fracture was launched to ensure the capability of the available chamber to withstand high gauge pressures up to 5 bar. Two types of design against fraction are presented here: the first one is a whole pressure vessel design study to determine the minimum wall thickness of a pressure vessel of a specific volume to withstand a certain maximum pressure, the second one is a detailed analysis against fracture for different weak components of the vessel like glass miniatures, bolts, and nuts.

## A.1 Whole Pressure Vessel Design

As described before, the used pressure chamber is of a cuboidal shape of internal length 0.37 m and made of 14 mm thickness steel plates St33 according to DIN 17100 Grade classification. According to ASME design code (BPVC VIII-1) [50] for designing pressure vessels (PV), there are two available design approaches for a pressure vessel as a cylinder or a sphere.

### A.1.1  Cylindrical design approach

In this calculation procedure, the used pressure chamber, which is of a rectangular cross-section, is first assumed to be in a cylindrical shape to get use of the available design code. Then, the design parameters are calculated and finally the real design parameters are expected using an assumed large factor of safety (FS) as the PV cross-section is not a cylindrical one.

As a cylinder, the following equations can be used to determine the required minimum thickness to withstand a maximum design pressure inside the chamber:

$$t_v = \frac{P \, D_i}{2 \, (S_D - P)} \tag{A.1}$$

Where, $S_D$ is the material allowable strength, P is the maximum design pressure inside the vessel, $D_i$ and $t_v$ are the internal diameter and thickness of the vessel, respectively.

Design stress based on ultimate strength ($S_U$) is calculated as:

$$S_D = 0.25 \, S_U \tag{A.2}$$

$$S_D = 0.25 \times 290 = 72.5 \text{ MPa}$$

Design thickness based on ultimate strength (for: P = 0.8 MPa, Di = 0.37 m) equals to:

$t_v$ = 2 mm, by applying safety factor of 4

$t_v$ = 8 mm.

Design stress based on yield strength ($S_Y$) is calculated as:

$$S_D = 0.625\, S_Y \qquad\qquad (A.3)$$

$S_D$ = 0.625 x 185 = 115.6   MPa

Design thickness based on yield strength (for: P = 0.8 MPa, Di = 0.37 m) equals to:

$t_v$ = 1.3 mm, by applying safety factor of 4

$t_v$ = 5.2 mm.

According to the given thickness of 14 mm, the permissible maximum pressure inside the vessel can be calculated by:

$$P = \frac{S_D}{\left(\dfrac{D_i}{2\,t_v} + 1\right)} \qquad\qquad (A.4)$$

The permissible maximum pressure based on ultimate strength

P = 51 bar, by applying safety factor of 4

P = 12.75 bar.

Moreover, the design should consider different primary stresses (either normal or shear stresses) which can lead to pressure vessel failure. The value of primary stresses depends on balancing all external loads including pressure, bending moment, torsional moment and shear forces. When PV thickness exercises yielding, PV wall thickness will be deformed, and strains are developed giving rise to the resisting internal stresses through stain hardening.

In the current configuration the test cell, which is placed inside the pressure chamber, is subjected to temperature gradients depending on the temperatures of both upper and lower walls of the test cell as described before. This leads to an increase in the mean temperature of the air, surrounding the test cell inside the pressure chamber. Therefore, according to ASME (BPVC VIII-1), the permissible maximum pressure inside PV is affected by wall deformations caused by temperature difference between the gas inside the PV and atmosphere, whether the pressurized gas is heated or cooled (emergency conditions). It depends on the vessel cross-section shape factors ($\alpha_o$ and $\beta$), where $\alpha_o$ is the ratio between full Plastic load and Initial yielding load.

$$\beta = 1 / \alpha_o \tag{A.5}$$

The new permissible vessel maximum pressure ($P_3$) can be calculated as:

$$P_3 = \frac{Spm3}{(\frac{D_i}{2\,t_v} + 1)} \tag{A.6}$$

Where Spm3 is the primary membrane stress (Axial, pressure stresses).

$$(Spm3 + Spb3) = - [(\alpha_o - \beta) / (1 - \beta)]\, Spm3 + [(\alpha_o - \beta) / (1 - \beta)]\, S \tag{A.7}$$

$$Spb3 = \tfrac{1}{2} (a \times E \times \Delta T) \tag{A.8}$$

Where Spb3 is primary bending stress, S is the design stress, a is coefficient of thermal expansion, E is elastic constant, and $\Delta T$ is the temperature difference between the gas inside the vessel and the atmosphere.

According to design stress based on yield strength, assumed temperature difference of 200 k, and for a pressure chamber of a rectangular cross section, then:

For a rectangular cross-section, $\alpha_o = 1.5$ and $\beta = 0.67$.

$S = S_Y = 185$ Mpa, $a = 10^{-6}$ mm/mm °C, and $E = 21000$ MPa.

From equation A.8(A.8, Spb3 = 252 MPa.

From equation A.7, Spm3 = 60.14 MPa.

From equation A.6, $P_3 = 2.19$ MPa = 21.9 bar.

Therefore, the current pressure chamber can withstand an internal gas pressure of about 20 bar when subjected to a temperature difference of 200 k, which is far from our operating pressure and temperature conditions.

### A.1.2  Spherical design approach

In this calculation procedure, the used pressure chamber, which is of a rectangular cross-section, is first assumed to be in a spherical shape to get use of the available design code. Then, the design parameters are calculated and finally the real design parameters are expected using an assumed large factor of safety as the PV cross-section is not a circular one.

The following equations can be used to determine the required minimum mass of a spherical pressure vessel (M) to withstand a maximum design pressure inside the vessel (P):

$$M = \frac{3\,P\,V\,\rho}{2\,\sigma} \qquad\qquad (A.9)$$

Where V is volume of the vessel, $\rho$ is material (Steel) density, and $\sigma$ is the design stress.

$V = \frac{4}{3}\pi x\,(\frac{0.37}{2})^3$ m$^3$, P = 20 bar, $\rho$ = 7750-8050 kg/m$^3$

According to a design stress based on ultimate strength and using a safety factor of 4, then:

$\sigma$ = 290/4 =72.5 MPa and M = 9 kg

By applying an approximation factor of safety of 4 (as an approximation from spherical shape of the design concept to the cuboidal real shape), then:

M = 36 kg, which is too small compared to the actual mass of the used pressure chamber.

From the Whole PV design criteria, it is easily concluded that the current pressure vessel can maintain a gas pressure of 8 bar inside it and can be also subjected to a temperature difference (between the gas inside the PV and the atmosphere) of 200 °C without any damage or fracture possibility.

## A.2 Detailed Design against Fracture for Some Components

The whole design against fracture method is usually sufficient to design normal pressure vessels. Unlike normal pressure vessels, the used pressure chamber contains different parts like sight glass elements, flanges, and bolts. Therefore, an additional design against fracture should be done for every single element with great considerable care to ensure to which value of gauge pressure all parts can withstand. The weakest parts in the pressure chamber described before are the glass elements and the bolts of both main steel plates and the small flange of the side glass element. The following study shows that all the parts can withstand at least an operating gas pressure of 4 bar inside the pressure chamber.

### A.2.1  Large front sight glass element

Figure A- 1 shows the front sight glass element that consists of two steel flanges, where a transparent glass of type ILMADUR® DIN 7080-8 is mounted. The glass element is of 250 mm diameter and 25 mm thickness, which can withstand till 8 bar as an operating pressure according to the manufacturer manual [51].

*Figure A- 1 Front glass element.*

### A.2.2  Small side sight glass element

Figure A- 2 shows the side sight glass element that consists of two steel flanges where a transparent glass of 87 mm diameter and 8 mm thickness is installed. According to the maximum permissible pressure for different glass plates found in [52], the allowed maximum pressure of the used glass plate can be estimated as the following:

$$s \geq 0.55 \, d_m \sqrt{\frac{P\,FS}{10\,\sigma_{DV}}} \qquad\qquad (A.10)$$

Where s is the theoretical minimum glass thickness in mm, $d_m$ is the diameter of glass plate in mm, P is the allowable pressure in bar, $\sigma_{DV}$ is the surface tensile stress in N/mm$^2$, and FS is factor of safety.

*Figure A- 2 Side glass element.*

According to equation A.10(A.10 , $d_m{}^2$ is inversely proportional to P. As found in [52], a glass plate of 63 mm diameter and 8 mm thickness can withstand a maximum pressure of 8 bar. Therefore, a glass plate of 87 mm diameter and 8 mm thickness can withstand a maximum pressure of 4.2 bar. On the same side, $s^2$ is directly proportional to P. As mentioned in [52], a glass plate of 90 mm diameter and 10 mm thickness can withstand a maximum pressure of 8 bar. Therefore, a glass plate of 87 mm diameter and 8 mm thickness can withstand a maximum pressure of 6.4 bar. Therefore, considering the worst-case scenario, the current used glass plate can withstand a maximum pressure of 4 bar.

### A.2.3  Small flange bolts

The side small glass plate is mounted through a flange of 130 mm diameter, which is fixed to the side steel plate of the vessel by means of four bolts. These bolts are from type A270 M8 with maximum allowable tensile strength of 700 N / mm$^2$. Assuming a maximum allowable pressure inside the vessel of P = 8 bar, the load on each bolt can be calculated as follows:

$$F_{bolt} = (P \times A_{flange}) \,/\, N \qquad\qquad (A.11)$$

Where $F_{bolt}$ is the applied force on each bolt and N is the number of bolts.

Then, $F_{bolt}$ = 2654.64 N and the actual tensile stress on each bolt $\sigma_{bolt}$ can be calculated using equation A.12.

$$\sigma_{bolt} = F/A_{bolt} \qquad\qquad (A.12)$$

$\sigma_{bolt}$ = 52.82 MPa.

Therefore, the tensile strength of the used bolts is 13 times more than the actual tensile stress load in case of an applied pressure of 8 bar inside the used pressure chamber. That ensures a higher degree of safety.

### A.2.4  Main steel plate bolts

Each steel plate is mounted to the main body of the vessel by means of eight bolts These bolts are from type A270 M16 with maximum allowable tensile strength of 700 N / mm$^2$. Assuming a maximum allowable pressure inside the vessel of P = 8 bar, the load on each bolt can be calculated using equations A.11 and A.12 as follows:

$F_{bolt}$ = 10752.1 N and $\sigma_{bolt}$ = 53.47 MPa.

Therefore, the tensile strength of the used bolts is 13 times more than the actual tensile stress load in case of an applied pressure of 8 bar inside the used pressure chamber.

www.ingramcontent.com/pod-product-compliance
Lightning Source LLC
LaVergne TN
LVHW041728190726
843493LV00007B/2252